Edward A Rix, A. E Chodzko

A Practical Treatise on Compressed Air and Pneumatic Machinery

Edward A Rix, A. E Chodzko

A Practical Treatise on Compressed Air and Pneumatic Machinery

ISBN/EAN: 9783337002640

Printed in Europe, USA, Canada, Australia, Japan

Cover: Foto ©berggeist007 / pixelio.de

More available books at **www.hansebooks.com**

A PRACTICAL TREATISE

ON

COMPRESSED AIR

AND

PNEUMATIC MACHINERY

BY

EDWARD A. RIX AND A. E. CHODZKO

PNEUMATIC ENGINEERS

FOR THE

———FULTON

Engineering and Shipbuilding Works

SAN FRANCISCO

———

MANUFACTURERS OF

MINING, MILLING, SMELTING AND
ELECTRICAL MACHINERY

ENGINES, BOILERS, HEATERS, PUMPS, ETC.

———

MAIN OFFICE AND BRANCH WORKS, 213 FIRST ST.
MAIN WORKS, HARBOR VIEW

SAN FRANCISCO, CALIFORNIA

1896

Curves, Tables and Engineering Data in the body
of this Treatise are original and were
prepared by

EDWARD A. RIX AND A. E. CHODZKO
PNEUMATIC ENGINEERS
San Francisco, - California

Entered according to Act of Congress, in the year 1896,
by
THE FULTON ENGINEERING AND SHIPBUILDING WORKS
AND
EDWARD A. RIX
In the Office of the Librarian of Congress,
at Washington, D. C.

Press of the Hicks-Judd Co.
23 First Street, San Francisco

COMPRESSED AIR.

It is a noteworthy fact that, while compressed air has been known and been used at a time when dynamic electricity was not even in its infancy, its properties and possibilities are still, in the minds of many practical people, an object shrouded with confusion and mystery, and considered by them as a convenient topic for the scientist's investigation, but altogether too intricate and obscure to be readily grasped by a man possessed of a common and average knowledge of motive machinery.

This same man, strange to say, will find no apparent mystery in handling a first-class Compound Condensing Steam Engine, whose thorough comprehension, however, involves a more imposing array of natural phenomena than does the action of an air motor.

Mention to him this latter machine, and he will tell you at once that it is useless; he has a vague recollection that compressed air will not yield over 15 to 20 per cent of the power expended to produce it, while an electric motor utilizes 60 or 80 per cent of this power, and that is the end of it.

The fact is, however, without in any way disparaging the wonderful strides made by electricity, that, in a great many circumstances, a compressed air power transmission will be found fully as much, and often more effective than an electrical transmission.

Within a radius of 10 to 20 miles or more, it is not a matter of theoretical speculation, but a result of actual facts, extending over a period of many years' experience, that compressed air can be economically produced, conveyed, and utilized as a motive power; and if this power is to be distributed throughout a number of buildings or factories, or in the interior of a mine, the absolute safety consistent with the use of compressed air is an element of superiority to which the electrical transmission has no possible claim.

However well insulated the conductors may be, the vicinity of a dynamo is always dangerous, either on the ground of fire or of bodily injury.

In a large power station, manned by a picked staff of attendants, this danger is small indeed; but the conditions are altogether different if the motor is under the care of a miner or of an ordinary workman.

Again, the location of an air motor is privileged with a constantly renewed and wholesome atmosphere, whose temperature can be, at will, regulated to suit the local exigencies.

Accidental circumstances which may occur in the vicinity of an electric wire under high potential are generally fraught with peril. The only accident to which an air pipe is liable is a leak, which will cause a loss of power, but which can be repaired and approached at no risk whatever.

But now comes another point.

Referring more especially to the mines which, in California, should represent a large percentage of the users of compressed air, an example will well illustrate the comparative merits of the two modes of power transmission, especially for mines.

Take a mine which was equipped some years ago; ample water power exists several miles away, but the configuration of the ground did not permit of conveying the water to the mine; a telodynamic transmission would not have been practical, so the owners concluded to put up a first-class steam plant for hoisting, pumping, and milling purposes, and also for running compressors supplying air to the rock drills, and perhaps to one or more underground pumps and fans.

In the course of time, however, timber has grown scarce in the surrounding sections, and now they have to haul their firewood at the rate of $5.00 or $6.00 a cord; as there is quite an amount of power used at the mine, this represents a rather burdensome item, so the owners begin to investigate some possible way out of it.

An electrical transmission is forthwith proposed to them, with tangential wheels and generators near the waterfall, conductors readily spanning all the intervening ridges and canyons, and a number of dynamos to replace the steam engines.

It is a practical, feasible, and satisfactory proposition, but there is one black cloud in this bright sky: what shall become of the steam engines and boilers? They have to be torn down, of course, to make room for the dynamos. This whole plant is still, however, in perfect condition; it has been bought, hauled, and erected at great cost, and would bring, at a sale, about as much as its equivalent of scrap iron, supposing it could be sold at all. The proposed plant will assuredly be more economical, but this is a dead loss which it will take some time to make up for.

Here comes the opportunity of the compressed-air man; he proposes to put up an air-compressing plant at the water-fall; the iron-pipe that carries the air will span ridges and canyons so easily, for all practical purposes, as did the wires, but after it reaches the mine, the list of new material closes, or nearly so; for neither the engines nor the boilers will have to be touched. The former will work with air as they did before with steam, the boilers being used for heaters or air receivers.

The compressor that used to work at the mine will not even have to be discarded, as it may serve either as a reserve or as a pressure transformer; in other words, there will have been an addition to the mine's possessions, in the shape of the compressors at the power-house and of the pipes, but the old plant will remain just as it was, and give full value for what it did cost; it will simply be necessary to find a new job for the firemen and woodchoppers.

Another very important point: suppose the power plant at

the water-fall met with accident, or the conductors to be temporarily crippled; with the electric plant it means a stoppage of the whole mine; with the compressed air proposition it would only be necessary to fill the boilers, start up the fires, and run by steam again. Here, there is no possible competition between the two systems. The advocates for electricity claim a superior economy, but a few developments on the production and the utilization of compressed air will, it is hoped, prove to the contrary, and we will try to illustrate the laws and properties of air and compressed air in a simple manner, and with the constant remembrance that practical men want plain facts and have no use for mathematical-discussions.

It is a common feature with gaseous substances that heat has a tendency to increase their volume, or, as the term goes, to expand them. Referring more particularly to atmospheric air, it will suffice to recall the classical experiment of the cork shutting hermetically a bottle full of air, and blown out, if the bottle be dipped in hot water.

Therefore, if a certain amount of air is confined within a closed cylinder, at the outside temperature, and then exposed to a source of heat, this air will have a tendency to expand, the result of which may be twofold.

If the cylinder is closed, for instance, by two covers tightly bolted on, and if its walls and covers are strong enough to resist deformation under this expansive tendency, the volume of air will remain constant, and its pressure will increase.

But if we suppose that one of the covers be removed, and replaced by a tight-fitting piston free to move in the cylinder, and loaded with a certain weight, when the air is at the outside temperature, the piston will descend in the cylinder until it is balanced by the pressure of the confined cushion of air.

If now the cylinder is heated, the piston will start slowly upward, and then stop when the expansion will have ceased; in this case, the load of the piston, and consequently the pressure of the air, have remained the same as before heating, but the volume of air has increased.

Summing up these simple facts, we will say, therefore, that the effect of heat upon this mass of air is, in the first case, *to increase its pressure under constant volume;* and in the second case, *to increase its volume under constant pressure.* The reverse would happen in both cases; i. e., if we take the closed cylinder full of hot air, and if we allow it to cool down to the outside temperature, the volume of this air will, of course, remain the same, but its pressure will fall gradually, until it becomes the same as it was before heating.

In a similar way, if we allow the cylinder with its piston to cool down to the outside temperature, the volume of air confined under the piston will shrink, and the piston will gradually drop down to the point where it was before the cylinder was heated, the pressure, of course, remaining constant.

Now, following this line of reasoning, we may conceive

that, if the temperature around the cylinder was made colder and colder, the pressure of the constant volume of air of the first case would keep dropping, and the volume of the mass of air at constant pressure, in the second case, would also keep shrinking, until, if such a process was carried on far enough, the mass of air which we have been considering would be condensed in volume to nothing, and have no pressure at all.

A simple calculation shows that such a result would occur at the temperature of 461 degrees below 0 Fahr., or 493 degrees below the freezing point of water.

This temperature, which has been approached, but never yet reached by any contrivance at present at our command, is, so far, a matter of mental conception, but we may, however, conceive its existence. It is called the *absolute zero*, and plays an important part in the study of the properties of gases.

The absolute zero is, therefore, *the temperature at which a mass of air would have neither volume nor pressure.*

Passing now to a seemingly different subject, although its close connection to the preceding facts will soon appear, a few words may be said about the fundamental principle which forms the basis of all questions relating to the mechanics of gases, the *Principle of Equivalence of Heat and Work.*

This principle, formulated in plain language, means that *whenever work is performed, it develops heat;* and conversely, that *whenever heat is generated, it can be transformed into work.*

The scope of this principle is exceedingly broad.

The elementary conception of work involves two distinct elements: a force and a motion, and the measure of the amount of work developed by a certain force is the product of this force, multiplied by its displacement.

Thus, if we exert a pull of 1 lb., and if we move 1 foot in the direction of this pull, the work that we have developed amounts to 1-foot pound.

But, while this definition is true in all cases, work, in natural phenomena, can assume a very great variety of forms, which, moreover, it is not necessary to enumerate here.

Our daily experience shows us some applications of this principle of equivalence, or correspondence, between heat and work.

That *work develops heat*, we can see in hammering a cold bar of iron, which soon becomes hot; we see it in the result of human exertion, in the heating of a shaft journal when the work of friction becomes too great; in the sparks showing at the contact of a revolving wheel, and of a brake-shoe, or at the periphery of a grindstone, etc.

That *heat can be transformed into work* has been shown in the preceding explanations, when we saw a weighted piston lifted by heating the air confined beneath it.

The steam engine is another indirect demonstration of the same fact; when the heat developed in the combustion of coal generates steam, which accomplishes some work on the piston of an engine.

It would be useless to multiply examples of this capital principle; suffice it to say, that whenever work is performed, there is a production of heat. This will not always be sensible, especially if the work is slow and gradual, because the heat is lost by radiation, by absorption in surrounding bodies, etc., as soon as it is developed.

This subject of the equivalence between heat and work has been exhaustively studied and verified, and it is now accepted as a fundamental axiom in mechanics.

One British Thermal Unit (B. T. U.) of heat, i. e., the quantity of heat required to raise by 1 degree Fahr. the temperature of 1 lb. of water, *corresponds to 778-foot lbs. of work.*

In other words, 778-foot lbs. of work applied to a certain mass of air, for instance, will develop in it 1 B. T. U. of heat; and conversely, an amount of heat of 1 B. T. U. stored up in this air can develop 778-foot lbs. of work.

The number 778, or coefficient of correspondence between heat and work, is known as the *Joule's Equivalent*, from the name of the physicist who first set precise rules in this respect. Joule had fixed the figure at 772-foot lbs., which was for years adopted as correct. Subsequent investigation led to make it 778, and the most recent developments put it at 779. In this treatise it has been taken as 778.

But it is now expedient to clearly explain how a certain mass of air, which has been subjected to work, and which has therefore accumulated a certain amount of heat, can conversely develop work corresponding to that heat.

Let us take a cylinder full of air at atmospheric pressure, and closed at one end, and then let us insert at the other end a piston in this cylinder, and exert an effort upon the piston; the air confined within the cylinder will be gradually compressed, and occupy a smaller volume. At the same time, its pressure will have increased, and this compression has absorbed a certain amount of work, which will be measured by the mean pressure which the piston has had to overcome, multiplied by the amount of its displacement.

The pressure on the piston represents a certain number of lbs.; the displacement represents a certain number of feet, and their product represents a certain number of foot-lbs., which measure the work of compression.

Suppose now that we release the piston; the air confined in the cylinder, and whose pressure was solely owing to the effort exerted on this piston, will immediately expand and push it back, and if there was no friction between it and the cylinder walls, it would resume its former position, when the air-cushion would be at atmospheric pressure again. In other words, every amount of work spent in compressing the air, would be entirely returned by the expansion of this air, or, *to any work of compression corresponds an equal work of expansion*, if these efforts follow each other instantly.

Here, we did not make any assumption as to the temperature of the confined air, which has been supposed to remain stationary. But now let us confine, with a piston, a certain amount of free air in a cylinder, and let us fix the piston in this position so as to prevent it from backing out; and, then, let us apply to the cylinder some source of heat.

The confined air will have a tendency to expand, and as the piston cannot move, the pressure will rise; if then we let the piston free, the confined air will push it out in expanding, until it resumes the atmospheric pressure, and the outside temperature, and with the same restriction as regards frictional resistances.

We see that in both instances there has been some expansive work done, and the force that produced it was supplied in the first case by the work of compression, and in the second case, by the heating of the air. We see also that in this latter instance, the pressure of air in the cylinder depended upon the amount of heat supplied to it, or, in other words, upon its temperature, and so did the expansion work.

Returning now to the definition of the absolute zero, as given, which marks, so to say, the ideal limit of existence of a gas so far as volume and pressure are concerned, we can readily conceive that 1 lb. of atmospheric air, at 60 degrees Fahr., for instance, is the outcome of 1 lb. of air at the temperature of absolute zero, to which a sufficient amount of heat has been supplied to raise its temperature by $461+60=521$ degrees Fahr., and its pressure to 14.7 lbs. per square inch, above a vacuum, which is the pressure at the absolute zero.

This pound of air is confined within the atmosphere, as was the mass of air of the last example within a cylinder; but should it be allowed to expand against a perfect vacuum, it would produce an amount of expansion work corresponding to the amount of heat which it had received to become atmospheric air.

This capacity of producing expansion work is what is termed the *Intrinsic energy* of this pound of air, and its existence is, as we see, intimately connected with the conception of the absolute zero.

The amount of work that measures this intrinsic energy can be determined from the law of the equivalence of heat and work, since we know that by storing up a certain quantity of heat in a mass of air, we give it the property of returning a corresponding quantity of work.

The temperature to which a given amount of heat will raise 1 lb. of different substances is not the same for all of them.

The *specific heat* of a substance is the number of B. T. U. that will raise by 1 degree Fahr. the temperature of 1 lb. of this substance, the specific heat of water being taken as unit. We have seen already that the specific heat of water was 1; i. e., that it takes 1 B. T. U. to raise by 1 degree Fahrenheit the temperature of 1 lb. of water.

The specific heat of air which we have to use in the subsequent developments is 0.2377.

In other words, it takes 0.2377 of a B. T. U. to raise by 1

degree the temperature of 1 lb. of air, that is to say, the amount of heat that would raise by 1 degree Fahr. the temperature of 1 lb. of water, will raise by 1 degree Fahr. the temperature of 4.2 lbs. of air.

The quantity of heat necessary to raise by 521 degrees Fahr. the temperature of 1 lb. of air is, therefore:

$$0.2377 \times 521 = 123.8412 \text{ B. T. U.},$$

and the corresponding amount of work is,

$$123.8412 \times 778 = 96,348.52 \text{ foot lbs.},$$

which represents the Intrinsic energy of 1 lb. of air at 60 degrees Fahr.

This, of course, presumes that no heat would be either lost or gained, by radiation or otherwise, during the expansion of air, and this sort of expansion is called *Adiabatic* expansion.

Now, while any one will readily understand that the expansion of air can be utilized to do useful work on a piston, it is also obvious, for practical reasons, that this expansion cannot be carried below atmospheric pressure, since creating a vacuum would require additional work.

Consequently, we cannot expect to avail ourselves of any portion of the intrinsic energy stored up in atmospheric air, under ordinary circumstances.

With a steam engine we can obtain a vacuum, or at least a pressure inferior to the atmosphere, by condensing the steam, but there is no such thing in the air machine.

Let us observe, moreover, that the intrinsic energy possessed by 1 lb. of air is entirely independent of its pressure, so long as its temperature remains the same, the work of expansion being exclusively controlled by the extreme temperatures between which the air expands; so that 1 lb. of air at 100 lbs. gauge pressure, and 1 lb. of air at 10 lbs. gauge pressure, and both at 60 degrees Fahr., possess the same total intrinsic energy as 1 lb. of atmospheric air.

But there is a vast difference between them at a practical standpoint, inasmuch as air at 100 lbs., and even at 10 lbs., can do some useful work by expanding down to atmospheric pressure; part of their intrinsic energy can, therefore, be utilized to do some actual work.

Taking, for instance, 1 lb. of air at 100 lbs. gauge, and at 60 degrees Fahr.—if allowed to expand adiabatically to atmospheric pressure, it will produce work, and consequently lose part of its heat, and we find that its temperature, after the expansion has taken place, is: — 173.95 degrees Fahr.

The drop of temperature is:

$$173.95 + 60 = 233.95 \text{ degrees.}$$

and as $778 \times 0.2377 = 184.93$, the work of adiabatic expansion is:

$$184.93 \times 233.95 = \qquad 43,264.37 \text{ ft. lbs.}$$

this being the *useful work*,

The adiabatic work of expansion from 173.95 degrees Fahrenheit to the absolute 0 would
be: $184.93 \times 287.05 = \qquad 53,084.15$ " "

Total, 96,348.52 " '

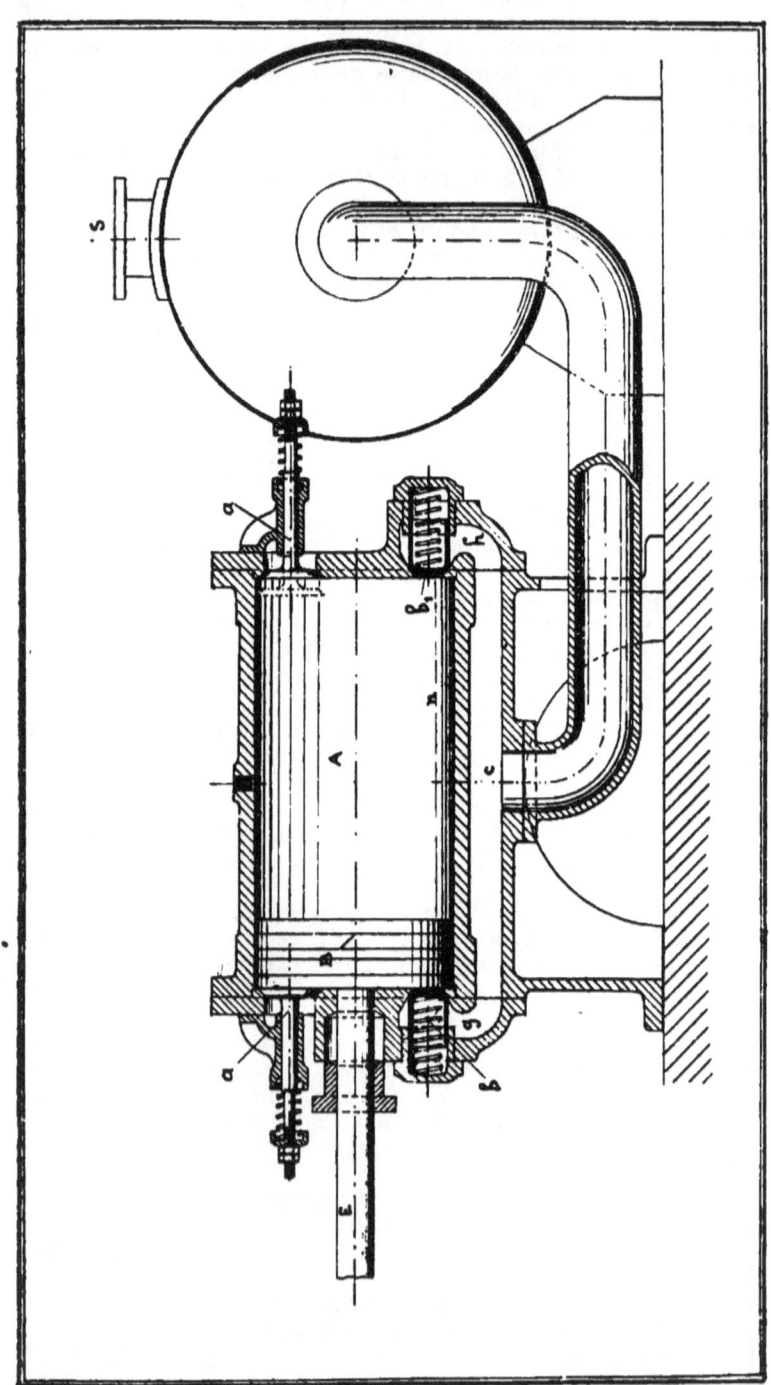

Fig. 1.—Diagram Illustrating Principle of Air Compression.

which is the total intrinsic energy—that is to say, we have utilized 45 per cent of the total intrinsic energy.

Next, taking air at 10 lbs. gauge, the temperature after adiabatic expansion to atmospheric pressure is — 12.9 degrees Fahr., and the useful work of expansion is:

$184.93 \times 72.9 =$ 13,481.39 ft. lbs.

The adiabatic expansion from — 12.9 degrees to absolute zero would give:

$184.93 \times 448.1 =$ 82,867.13 " "

Total, 96,348.52 " "

i. e., the total intrinsic energy, and the useful work is here 14 per cent of the total intrinsic energy.

It is hardly necessary to say that these figures are theoretical, because, in practise, part of the work of expansion, and consequently part of the heat, is absorbed by the friction of the piston in the cylinder, and lost by radiation from the various pieces of the machines.

We see, therefore, that the only portion of the intrinsic energy of air that is practically obtainable is the expansion work which it does above atmospheric pressure; i. e., that the pressure of this air must be raised above the pressure of the atmosphere.

From the preceding developments we might rightly conclude that this result would be reached by heating the air, previously confined within a closed vessel, to a proper temperature. But in practise, such a process would prove unacceptable.

Compressed air is slow in taking up heat, because its conductivity is small; i. e., because the heat is slow to penetrate the whole mass of air, and its low specific heat causes it to cool down rapidly.

Then, again, the whole amount of expansive work above atmospheric pressure could not, as said before, be obtained in practise; so that raising the pressure of air by mere heating is not a practical proposition, and it is necessary, in order to meet the requirements of its industrial applications, to operate this rise of pressure by direct compression; i. e., by acting upon the air, confined in a cylinder, through a piston to which an adequate amount of power is applied.

This compression, in whichever way the rise of pressure occurs during its process, is always affected on the following general lines:

A cylinder A (Fig. 1), closed at both ends by covers, contains a piston B, which can move back and forth therein, and whose rod C is connected, either to the piston of a steam engine, or, through a connecting-rod and a crank, to a revolving shaft.

Each one of the cylinder covers carries one or more inlet valves a, a', through which the atmospheric air can penetrate into the cylinder; each valve, of course, opening inward, and being maintained tightly pressed upon its seat by a spring.

The covers also carry one or more discharge valves C C', similarly kept closed by a spring, and opening outward into closed chambers g, h, connected by a common conduit c, which leads to a closed receiver r, whence a pipe attached to the nozzle s, conveys the air to the place where it is proposed to use it.

All the valves being closed, and the piston B at one end of its stroke, as shown, if it is set in motion from the left to the right, a partial and increasing fall of air pressure will occur behind it, and soon overcome the tension of the spring which keeps the inlet valve a closed; this valve opens, and atmospheric air rushes into the cylinder, behind the receding piston.

On the right side of this latter, we have, at the beginning of the stroke, a cylinder full of atmospheric, or, as generally called, of free air; the inlet valve a, and discharge valve b, are both closed, and so remain as the piston moves from left to right, because the air pressure in the cylinder has a tendency to close the inlet valve a', whilst its pressure is not sufficient to lift the discharge valve b'.

The piston continuing to move, the air pressure constantly increases, until, at a certain point n of the stroke it reaches, or slightly surpasses, the receiver pressure.

The action of this latter on the outerside of the discharge valve b', and also the tension of its spring are now balanced, and the smallest subsequent move of the piston opens this valve, and the compressed air is forced through it into the receiver, until the piston reaches the end of its stroke, when the discharge valve is closed by its spring.

An inverse series of operations will occur during the reverse stroke, and so on.

An analysis of these operations shows that during any one stroke of the piston there are three distinct classes of work performed: on one side of the piston, a work of suction; on the other side, first a work of compression, under variable piston load, and then a work of delivery, under constant piston load.

This is quite similar, only in the reverse order, to what occurs in the cylinder of a steam engine, wherein a certain volume of steam is admitted under full pressure, and then, after cutting off its ingress, is allowed to expand during the remainder of the stroke.

The work of suction, which overcomes the inertia of the inlet valves, the tension of their springs, and the resistance of air in its passage through the valve apertures, is always small, and can be reduced by properly proportioning and constructing the inlet valves.

It is, therefore, a matter of correct design, which has nothing to do in the present developments, and no further mention of it will hereafter be made.

Of the two other qualities of work, the period of delivery does not either offer any peculiar feature to investigation besides its relative proportion to the whole stroke, inasmuch as it is symbolized by a constant load acting against the piston, along a certain distance, which corresponds to the elementary definition of work as previously given.

We are thus left to concentrate our attention upon the period of compression.

The variations of volume and of pressure of air, which occur gradually during the process of compression, do not follow the same law in all cases; that is to say, this variation is different, whether the compression takes place at a constant temperature (isothermal compression) without any loss or gain of heat, or by allowing the increasing heat developed during the compression to remain integrally in the air; in other words, if the compression is done at variable temperature (adiabatic compression).

There is no intention to develop here the laws governing the pressure and volume of air in those two sorts of compression.

This would necessarily involve the use of mathematical formulæ, which we wish to avoid. Suffice it to say that, *if the temperature of the air remained constant throughout the compression*, the volume which it occupies at any moment would vary inversely as the pressure.

Taking, for instance, 1 cubic foot of free air at 60 degrees Fahr., its pressure is, therefore, 1 atmosphere, or 14.7 lbs. per square inch above a vacuum, or also zero gauge pressure. Suppose that this air is confined under the piston of a closed cylinder, and that, driving this piston forward, we reduce the volume occupied by the air to ½ cubic foot only, at the same time maintaining always its temperature at 60 degrees Fahr. Then the pressure of this air would be 29.4 lbs. per square inch above a vacuum (or 14.7 lbs. gauge), that is, twice what it was before.

If the volume was reduced to ⅓ of a cubic foot, its pressure would become 3 × 14.7, or 44.1 lbs. per square inch above a vacuum or 29.4 lbs. gauge, always upon the condition that the temperature remains, throughout this process, at 60 degrees Fahr.

In other words, if the volume of air becomes 4, 5, 6, 10, 20 times smaller, its pressure becomes 4, 5, 6, 10, 20 times greater, *always taking the pressure of the atmosphere* (or the gauge pressure plus 14 7 lbs. per square inch) *as unit*, and not the gauge pressure, which would lead to absurd conclusions.

These pressures counted above a vacuum are called *absolute pressures;* the pressures indicated by the pressure gauge of a boiler are termed *effective* or *gauge pressures.* The absolute pressure is obtained by adding 14.7 lbs. to the corresponding gauge pressure; and conversely, the gauge pressure is obtained by subtracting 14.7 lbs. from the corresponding absolute pressure.

Let us take a cylinder open at one end (Fig. 2) and a piston moving in it. Suppose that the piston is at 48 inches from the cylinder head, that this space has been filled with free air through the inlet valve, and that the pipe leading from the discharge valve casing communicates with a receiver wherein the pressure is 73.5 lbs. gauge per square inch.

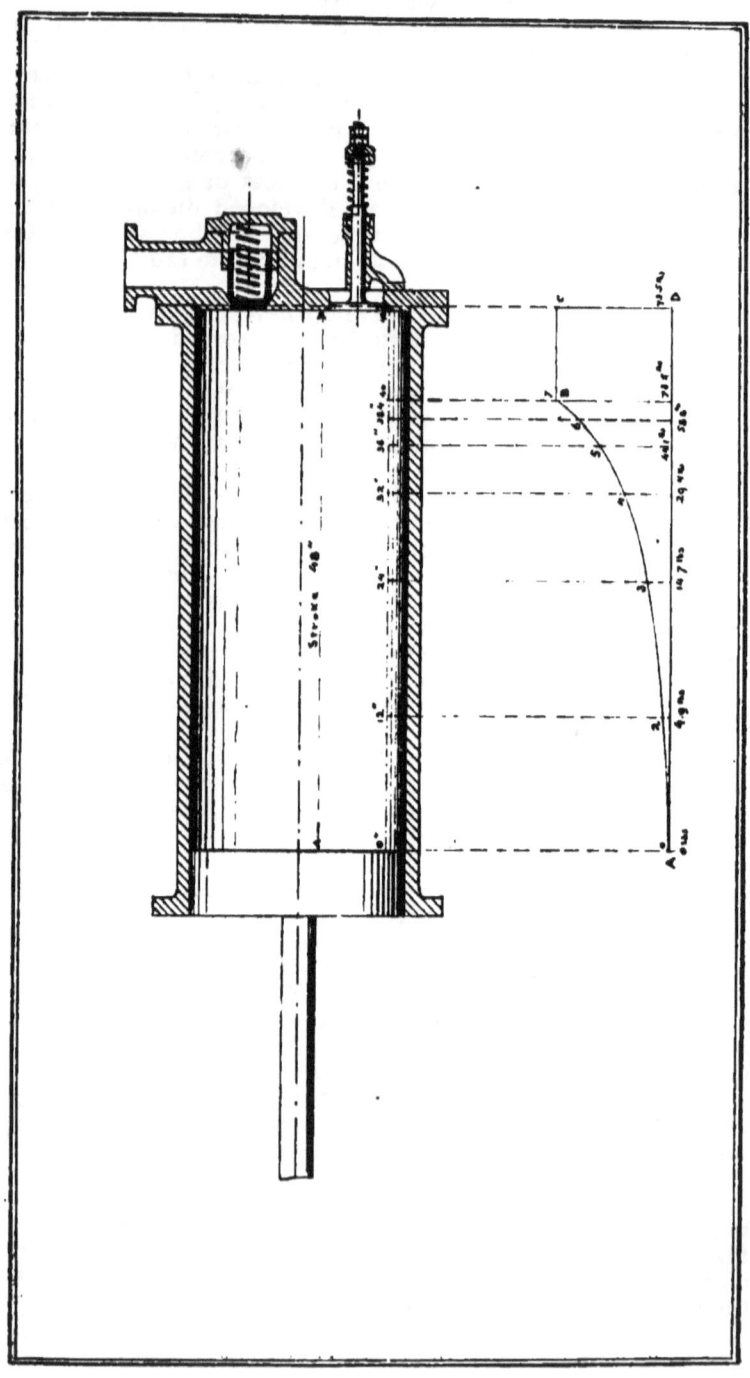

We will assume, also, that the compression is isothermal; i. e., that the temperature in the interior of the cylinder remains the same as in the open air.

If we move the piston 12 inches, the volume occupied by the air is 36 inches, or $\frac{3}{4}$ of its former length, 48 inches. The pressure must, therefore, be the reverse, or $\frac{4}{3}$ of the atmospheric pressure; i. e., 19.6 lbs. absolute, or 4.9 lbs. gauge.

Similarly, when the piston has successively covered 24, 32, 36, 38.4 46 inches of its stroke, the absolute pressures are respectively:

29.4, 44.1, 58.8, 73.5, 82.2 lbs., and the gauge pressures 14.7, 29.4, 44.1, 58.8, 73 5 lbs., per square inch, which are marked on the sketch.

If the piston moves further on, as the pressure in the cylinder is the same as in the receiver, the discharge valve opens; there is no more compression, and the remaining 8 inches of stroke are completed by the piston against a constant gauge pressure of 73.5 lbs. per square inch.

Let us now draw a line, $A\,D$, which, at any scale, represents 48 inches, and mark on this line some points at 12, 24, 32, 36, 38.4, and 40 inches from its left end; then draw at those points some lines 12-2, 24-3, 32-4, 36-5, 38 4-6, 40-7, perpendicular to $A\,D$.

Now, on these lines, let us carry, at any other scale, the gauge pressure at the corresponding point of the stroke; this will give us a succession of points 2-3-4-5-6-7, and, if we join them by a continuous line, a curve $A\,B$, that represents the variations of air pressure during the compression.

This curve starts from the point A, where the gauge pressure is zero.

If we took any number of intermediate points between 40 and 48 inches of the stroke, the pressure would always be 73.5 lbs. gauge, and consequently the curve of compression $A\,B$ is followed by a line $B\,C$, parallel to $A\,D$, and representing the delivery under constant pressure: so the diagram $A\,B\,C\,D$ gives us a graphic representation of the isothermal compression and delivery of air during one stroke of the piston, and its area represents the work performed during that stroke, for each square inch of piston area.

The law of isothermal variation of the pressures and volumes applies to decreasing pressures as well as to increasing ones; thus, if we cause one cubic foot of air at 73.5 lbs. gauge (88.2 absolute) to occupy 6 cubic feet, its pressure will become 14.7 lbs. absolute per square inch or o gauge pressure.

In other words, the curve of isothermal compression is also the curve of isothermal expansion, and the diagram $A\,B\,C\,D$ represents either the work of compression and delivery of a volume of free air, to 73.5 lbs. gauge, or the expansive work of the same body of air at 73.5 lbs. gauge pressure, and expanded from that pressure to the atmosphere, when it resumes its prim-

certain mass of air in a closed cylinder, by pus forward by a certain number of inches, and th piston free, the air will expand and push it be there be no friction between the cylinder and latter would return exactly to its starting-pc during its reverse stroke exactly as much w required to push it forward.

Quite a similar course of reasoning leads us if we compress air isothermally in a cylinder, ar valve chamber (this valve being loaded to 1 communicates through a pipe of any lengt cylinder exactly alike, located at some distanc pressing cylinder, we can obtain isotherm resistances) from the second cylinder the s work that has been developed in the dista The first cylinder is *the compressor*, the seco connected by the air main to the compressor; perfect *compressed air transmission*, wherein a ; work is integrally conveyed to any distance f production.

But were we to establish such a system we in practise the work recovered from the motc equal to the work developed in the compresso

To reduce this difference (which is the key in this system of power transmission) to possible, constitutes, in a nutshell, the whole matic engineering; and as the first condi difficulty is to locate it, and to size it up, these concluded by a few explanations showing whe of discrepancy between the work expended in and the work recovered from the motor, and b partly eliminated; their total disappearance, o acting, being a purely practical matter, which limitations.

The fact of compressing air in a cylinder is panied by a production of heat. What cau develop in the case of air is a question the p which would carry us too far into theory. however, that modern science considers ai minute particles in a constant state of vibratio pressing a volume of air which contains a ce these particles causes them to increase the ; vibratory motions, hence friction, impact, and

Direct experiment, made from the freezin point of water, has shown that the pressure the same, its volume at 32 degrees Fahr. incre each increase of 1 degree Fahr. in the temper;

From this we see that air at the temper water has increased in volume by $^{180}/_{493}=0$ cent, whilst this same air, at 493 degrees be point of water, or 461 degrees below zero Fal by $^{493}/_{493}$ of its volume, or by that volume itse the temperature of absolute zero was ascertain

Compression will generate heat, and only should it be possible to eliminate it as soon as produced, would isothermal compression be obtainable. It might probably be done by a very slow and gradual compression, combined with copious means of cooling the air in the compressing cylinder.

But these conditions correspond to a practical impossibility, and there is in consequence a considerable amount of heat disengaged during the compression. The following table gives the temperatures Fahr. of dry air at the end of its compression, to different gauge pressures in adiabatic compression; i. e., supposing that no portion of the heat developing is lost in the course of compression.

Absolute Pressure. (Lbs. per sq. in.)	Gauge Pressure. (Lbs. per sq. in.)	Fahr. temperature at end of compression.
14.7	0	60°
16.17	1.47	74.6°
18.37	3.67	94.8°
22.05	7.35	124.9°
25.81	11.11	151.6°
29.4	14.7	175.8°
36.7	22	218.3°
44.1	29.4	255.1°
51.4	36.7	287.8°
58.8	44.1	317.4°
73.5	58.8	369.4°
88.2	73.5	414.5°
102.9	88.2	454.5°
117.6	102.9	490.6°
132.3	117.6	523.7°
147	132.3	554°
220.5	205.8	681°
294	279.3	781°
367.5	352.8	864°

We see that as the pressure increases so does the temperature, and that when, for instance, the pressure has reached 73.5 lbs. gauge per square inch, the temperature is 414.5 degrees Fahr., instead of 60 degrees, as was the case in isothermal compression.

The result is, that if we take the same cylinder which was used in that case, i. e., if we act on the same weight of free air, this air, when at 73.5 lbs. gauge, will be 354.5 degrees Fahr. warmer in adiabatic compression than it would in isothermal compression. Its volume must therefore be necessarily greater in the former case, since the pressure is supposed to be the same.

The practical meaning of it is that in adiabatic work the period of compression is shorter and the period of delivery is longer than in isothermal work; as the work at full pressure is naturally greater than at any time during compression, when the pressure is smaller, the adiabatic work is greater than the isothermal work, to raise the same weight of air to the same pressure.

For 73.5 lbs. gauge, and atmosphere at 60 degrees Fahr.,

the adiabatic work is 1.31 times the isothermal work. But if the work done by the motor is correspondingly greater, what harm does the heat do? There would be none but for the fact that the motor is always at some distance from the compressor (otherwise there would be no reason to transmit power), and the air parting easily with its heat, its passage through the receiver and the main will reduce the air to the temperature of the atmosphere; i. e., after compressing a volume of hot air $F\ C\ D\ G$ (Fig. 3), we shall introduce in the motor a volume $B\ C\ D\ E$ of cold air of the same weight and pressure.

Now, this volume will expand in the motor either isothermally or adiabatically.

As we saw that the work of compression disengages heat, similarly, but conversely, does the work of expansion absorb heat from the surrounding bodies, and as the isothermal compression would require a slow process with copious cooling, so would the isothermal expansion require a slow process with copious heating. Unless this is done, the expansion will be *rather* adiabatic.

Rather, because if isothermal conditions never strictly obtain in practise, the same is true with adiabatic work.

If we expand adiabatically the volume of air $B\ C\ D\ E$, at 60 degrees Fahr. and 73.5 gauge, to atmospheric pressure, the work of expansion represented by the diagram $B\ C\ D\ K$, will only be 0.595 of the work of adiabatic compression.

A compressed air transmission seems, therefore, to be an inferior system, the more so as the above figures do not take into account all the losses incurred, but only the thermic losses; i. e., such as are due to loss of heat.

Several means are resorted to in order to reduce this loss.

Suppose that the volume of cold air, $B\ C\ D\ E$, when it arrives at the motor, be reheated at constant pressure (73.5 lbs. g.) until it becomes equal to $F\ C\ D\ G$; then we shall be able to develop in the motor by the expansion of this volume of hot air the same work that was used to compress it adiabatically.

So if there was no other loss, the motor would utilize 100 per cent of the work of compression. Indeed, should the air arriving at the motor be reheated to a higher temperature than that reached in the compressor, the work recovered would be greater than the work expended; and there is no absurdity in this statement, for such a result is easily attained at the cost of a certain quantity of fuel, which must be taken into account and deducted in figuring up the actual efficiency of the motor.

Reheating the air upon its arrival at the motor is, indeed, the base of the superiority of compressed air as a medium of power transmission.

No corresponding feature exists with electricity to the possibility of increasing at any time the intrinsic energy of the motive agency in an easy and inexpensive manner.

There are, however—at least at present—some practical limitations to this reheating; compressed air cannot conve-

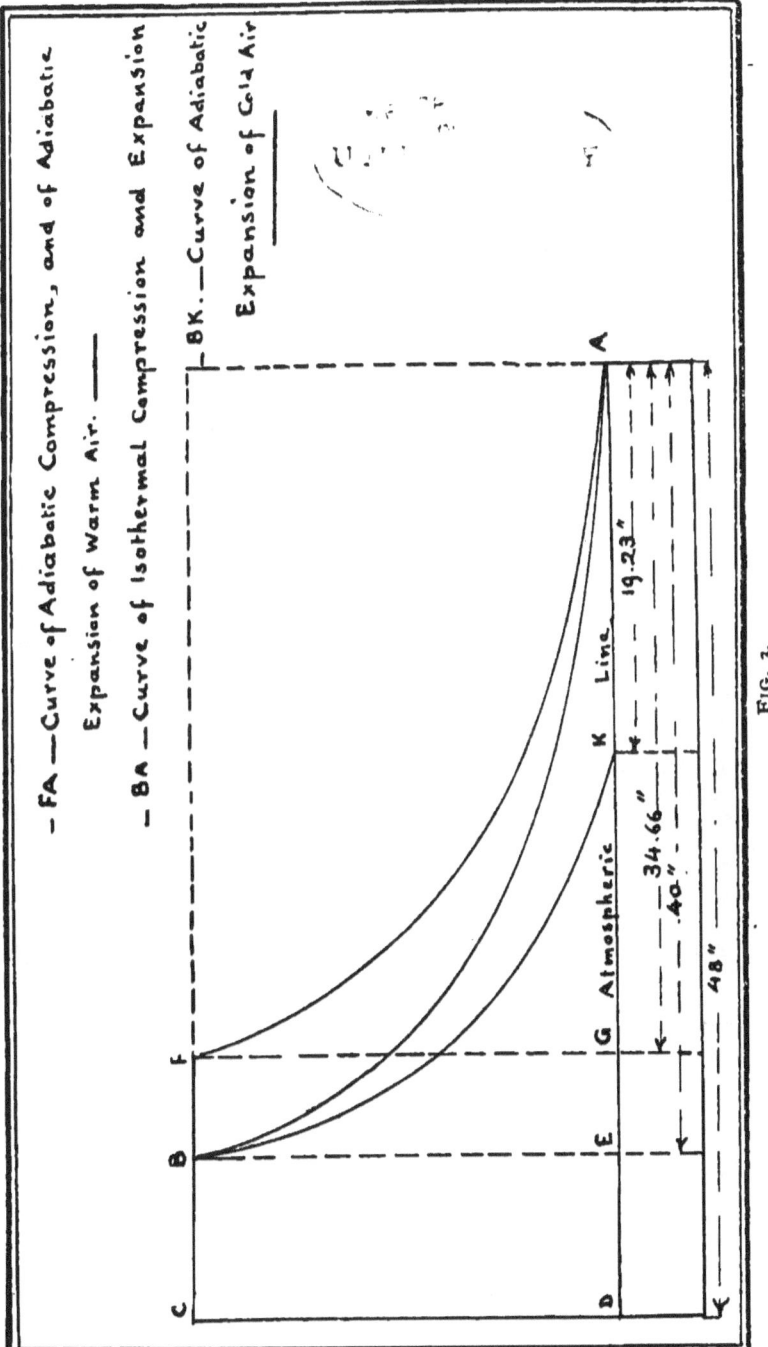

FIG. 3.

niently be admitted into a cylinder at a temperature much above 350 degrees Fahr.; while we have seen that in adiabatic compression, the temperature corresponding to 73.5 lbs. gauge is 414.5 degrees, and this illustration points out one reason why low pressure air is more economical, for power purposes, and also the use of compound compression where the rise of adiabatic temperature is small in comparison to single stage machines.

No lubricants of the ordinary description will be fully active beyond this temperature: special oils, however, are made which are not decomposed before 500 to 600 Fahr. But it is evident that, could the compressor and motor cylinders, pistons, and packings be made of a substance that would withstand great heat without injury or the usual lubricants, the reheating could be carried far enough to compensate for all other losses, a feature exclusively characteristic of air; and there is no apparent reason why such a substance could not be discovered, and it will reward undoubtedly its discoverer in a day not far distant.

Without entering into many particulars, it may be stated that when the compression from atmospheric to receiver pressure is effected in one cylinder, the air is cooled either by surrounding the walls of the cylinder with a jacket, and providing in the heads some hollow chambers, through which a continuous stream of cold water is rapidly circulated: this is called "dry cooling," because no water comes in contact with the air, and represents, without exception, the best American practise.

Or else, a spray of finely divided cold water is injected in the body of air under compression.

Here, the contact is direct between air and water, so this system is more effective than the dry cooling; and if the jacket arrangement is used in connection with the spray, a marked improvement occurs in the cooling of air.

This wet cooling has, however, some practical disadvantages, which led to discarding it in this country, while in Europe i. is still found in recent high-class compressing plants.

Another effective means of cooling consists in *compounding* the compressor; i. e., in effecting the total compression through a series of successive cylinders, in each one of which only a partial compression is effected, generating little heat, which is more easily dealt with; besides, the air in passing from a cylinder to the next one in the series is discharged through a cooler, where it resumes the outside temperature.

For high pressures, compounding is a necessity, and the efficiency of the compression is thereby increased.

The ideal compression is, of course, the isothermal; and its efficiency being 1, we have the following relative efficiencies for other systems, when air is compressed to 6 atm. effective, namely,

Adiabatic, without cooling 0.744
Adiabatic, with jackets.............. 80
Adiabatic with spray...................... 0.85

Adiabatic compound (2-stage), jacketed, with intercooler but no spray...... 0.863
3-Stage compound, with intercoolers and with spray...... 0.955

The effect of cooling is first to improve the efficiency of the compression; i. e., to use less work in producing it; and then, at the same time, the amount of heating is proportionally reduced.

A mere mention will be made of the loss incurred by the air pressure on its passage through the main connecting the compressor and the motor.

Tables will be found in this book giving the loss of pressure through mains of different lengths and sizes, and for different velocities of circulation of air.

In a general way, and especially in a long transmission, there is a conflict between the first cost of the plant and its efficiency, which both increase with the size of the main.

In a properly built line, the loss can be made very small, and its value will generally be assumed to suit local convenience, according to whether the first outlay or the cost of power is to have more weight in arriving at a decision.

It is hoped that the preceding remarks will enable any intelligent reader to form a correct idea of the elements to be taken into consideration in installing a compressed air plant or compressed air transmission, and briefly they may be enumerated:

1. An economical prime motor.
2. A compressor, which, while having a high mechanical efficiency, has also means for reducing the heat of compression to a minimum during and between the periods of compression.
3. A pipe line involving the least loss by friction consistent with the finances at command.
4. Motors, which, beside possessing a high mechanical efficiency, have means to expand the air to the atmospheric pressure, which must be done by reheating to as great a temperature as possible, both before and during the expansion of the air in the cylinders or upon the motor wheel.

TABLES FOR THE LOSS OF PRESSURE OF AIR IN PIPES.

In calculating tables for the loss of pressure in pipes, it has been found necessary to take a wide departure from the form of the tables usually given in catalogues on Compressed Air, and whose simplicity unfortunately does not agree with more recent experimental results touching upon the subject.

The formulæ from which such tables are generally established are the outcome of experiments made at the Mount Cenis Tunnel, and of Stockalper's more recent investigations at the St. Gothard Tunnel.

Similar formulæ have been used, with some modifications of detail, by Professor Riedler, who conducted extensive tests upon the compressed air system in Paris, and they are based on the assumption that the loss of pressure varies *directly as the length of the pipe*, and *inversely as its diameter*.

Professor Unwin took up the subject, availing himself of the results formerly obtained, and his investigation of the laws governing the motion of air in long pipes does not support the above-quoted conclusions.

Taking, for instance, three pipes, each 5 inches in diameter, wherein air enters at a pressure of 70 lbs. gauge, and at a velocity of 20 feet per second, if one of these pipes be one mile long, the second 2 miles, and the third 5 miles long, the loss of pressure according to Unwin's formula is:

4.6 lbs. for the 1-mile pipe,
9.4 lbs. for the 2-mile pipe,
26.3 lbs. for the 5-mile pipe.

In other words, the lengths being as: 1 – 2 – 5, the drop of pressure varies as: 1 – 2.043 – 5.72; and while the discrepancy is unimportant for short lengths, it becomes 14.4 per cent at 5 miles, and would be still greater for longer pipes.

As the logical tendency is toward increasing the practical length of power transmissions, a saving of a few pounds of loss is important; consequently, in working out a compressed air transmission, more precise data are needed. To meet this requirement, the following tables were calculated from Unwin's formula.

From the preceding example we notice that the loss of pressure increases more rapidly than the ratio of the lengths; besides, this loss does not vary inversely as the diameter of the pipe.

Taking a 4-inch pipe, 2000 feet long, into which air at 60 lbs. gauge enters with a velocity of 15 feet per second, the loss at the lower end will be *1.195* lbs.; according to the old rule, the loss in an 8-inch pipe of same length, and at the same pressure and velocity of air, would be one-half this amount, or *0.5975* lbs.; yet Unwin's rule makes it *0.52* lbs.

In the same way the loss in a 12-inch pipe should be *0.398* lbs., while its actual value is *0.3* lbs. Here the loss of pressure decreases more rapidly than the diameter increases.

And if we accept the theory that recent rules, when emanating from a reliable source, are the best, we must conclude that no satisfactory approximation to exact results can be obtained with the proportional formulæ.

In the annexed tables, the air pressure at the entrance to the main has been assumed to be 70, 80, 90, and 100 lbs. gauge, which figures cover the working pressures at which air will generally be admitted to the motors.

The use of the tables involves a few elementary operations, which we have clearly defined in several numerical examples, selected to suggest a ready method of solving any ordinary problems.

Some little calculation must of necessity be done, inasmuch as to construct a series of tables, which would take into consideration every element which influences all cases of transmission, would necessitate too much elaboration, and would not be desirable in a treatise of this character.

EXAMPLE 1.

500 cubic feet of free air is compressed per minute to 80 lbs. gauge, and conveyed through a 6¼-inch pipe, 2 miles long. What will be the air pressure at the lower end of the pipe?

Referring to Table Fig. 5, which deals with air compressed to 80 lbs. gauge, and starting down column 3 (size of pipe in inches) we stop at 6-¼ ins. On the left side (Col. I) we find for the ratio of absolute air pressures at lower and upper ends of main:

$$\sqrt{1-0.00000003709\ v_1^2\ l}$$

(v_1 is the velocity of air at entrance to main, in feet per second, and l is the length of pipe in feet.)

Following now the horizontal line to the right until it meets the vertical column headed 500, we find 36.5 which is the value of v_1^2.

So $v_1^2\ l = 36.5 \times 5280 \times 2 = 385,440$
and the ratio of air pressures (Col. I) becomes:

$$\sqrt{1-0.00000003709 \times 385,440} = 0.992$$

The pressure at entrance to main is 80 lbs. gauge or 94.7 lbs. absolute; the pressure at the lower end will be:

94.7 × 0.992 = 93.9 lbs. absolute
14.7
———
Or 79.2 lbs. gauge.

The loss is: 80 — 79.2 = 0.8 lbs.

EXAMPLE 2.

How many cubic feet of free air per minute, compressed to 90 lbs. gauge, can be conveyed in a 9-⅝ inch pipe, 5 miles long, the loss of pressure to be 3 lbs.?

The absolute pressure at entrance to main is: 104.7 lbs.
The absolute pressure at lower end is: 101.7 "

Their ratio is: $\dfrac{101.7}{104.7} = 0.971$

Referring to Table Fig. 6 (90 lbs. gauge) and following Col. 3 down to 9-⅝ inch, we find on the left of this figure (Col. 1)

— Ratio of — Absolute Pressure at Lower End of Main to Absolute Pressure at Entrance	Inside Diameter of Pipe		— Cubic Feet of Free Air Compressed per Minute to — 70 Lbs Gauge per Square Inch. —																	
	Feet	Inches	100	200	300	400	500	600	700	800	900	1000	1500	2000	2500	3000	3500	4000	4500	5000
$\sqrt{1 - .0000002045\, v_1^2 l}$	.166	2	174.5	698	1570.5	2792	4362.5	6282	8550.5	11168										
$\sqrt{1 - .0000001077\, v_1^2 l}$	.25	3	34.45	137.8	310	551.2	861.2	1240	1688	2204.8	2790.8	3435	7751	13780	21581					
$\sqrt{1 - .00000006858\, v_1^2 l}$	.33	4	10.95	43.8	98.6	175.2	273.8	394.2	536.6	700.8	887	1096	2464	4380	6844	9855	13414	17520	22179	27375
$\sqrt{1 - .00000004972\, v_1^2 l}$	.415	5	4.49	17.9	40.4	71.8	112.25	161.6	220	287.4	363.7	449	1010.2	1796	2806.2	4041	5500	7184	9092	11225
$\sqrt{1 - .00000003709\, v_1^2 l}$	.52	$6\frac{1}{4}$	1.83	7.3	16.5	29.3	45.8	65.9	90	117	148	183	412	732	1144	1647	2242	2928	3706	4575
$\sqrt{1 - .00000002990\, v_1^2 l}$	.6	$7\frac{1}{4}$	1	4	9	16	25	36	49	64	81	100	225	400	625	900	1225	1600	2025	2500
$\sqrt{1 - .00000002573\, v_1^2 l}$	.68	$8\frac{1}{4}$	.602	2.4	5.42	9.63	15.1	21.7	29.5	38.5	48.8	60.2	135.5	241	376.2	542	737.5	963.2	1219	1505
$\sqrt{1 - .00000002075\, v_1^2 l}$	.8	$9\frac{5}{8}$	.325	1.3	2.9	5.2	8.1	11.7	15.9	20.8	26.3	32.5	73.1	130	203	292.5	398	520	658	812.5
$\sqrt{1 - .00000001835\, v_1^2 l}$	.88	$10\frac{5}{8}$	.221	.884	1.99	3.54	5.53	7.96	10.8	14.1	17.9	22.1	49.7	88.4	138.1	199	270.7	353.6	447.5	552.5
$\sqrt{1 - .00000001636\, v_1^2 l}$	.96	$11\frac{5}{8}$	.1521	.61	1.37	2.4	3.8	5.5	7.45	9.7	12.3	15.2	34.2	60.8	95.1	136.9	186.3	243.4	308	380.25
$\sqrt{1 - .00000001466\, v_1^2 l}$	1.04	$12\frac{1}{2}$	.114	.456	1.03	2.12	2.9	4.1	5.6	7.3	9.23	11.4	25.7	45.6	71.33	102.6	139.7	182.4	230.8	285
$\sqrt{1 - .00000001362\, v_1^2 l}$	1.12	$13\frac{1}{2}$	.084	.34	76	1.34	2.1	3	4.12	5.4	6.8	8.4	18.9	33.6	52.5	75.6	102.9	134.4	170.1	210
1	2	3	4	5	6	7	8	9	10	11	12	13	14	15	16	17	18	19	20	21

— Values of V_1^2. —

— Remark — l is the length of Main in Feet — V_1 is the Velocity of Air at Entrance to Main in Feet per Second —

FIG. 4.—Friction Losses of Air in Pipes.

Ratio of — Absolute Pressure at Lower End of Main to Absolute Pressure at Entrance	Inside Diameter of Pipe		Cubic Feet of Free Air Compressed per Minute to 80 Lbs Gauge per Square Inch																
	Feet	Inches	100	200	300	400	500	600	700	800	900	1000	1500	2000	2500	3000	3500	4000	4500	5000
$\sqrt{1-.0000002045\,v_1^2\ell}$	.166	2	137.4	549.4	1236.6	2198.4	3435	4946.4	6732.6	8793.6	11129.4	13740	30915	—	—	—	—	—	—	—
$\sqrt{1-.000000107\,v_1^2\ell}$	.25	3	27.7	110.8	249.3	443.2	693.3	997.2	1357.3	1772.8	2243.7	2770	6232.5	11080	17312	24930	33940	44320	56115	69250
$\sqrt{1-.00000006858\,v_1^2\ell}$	.33	4	8.76	35.04	78.84	140.16	219	316.4	430.2	560.6	709.6	876	1971	3504	5475	7884	10731	14016	17744	21900
$\sqrt{1-.00000004972\,v_1^2\ell}$	.415	5	3.61	14.44	32.5	57.8	90.3	130	176.9	231	292.8	361	812.3	1444	2256.3	3249	4423	5776	7313	9025
$\sqrt{1-.00000003709\,v_1^2\ell}$	.52	6¼	1.46	5.84	13.14	23.4	36.5	52.6	71.5	93.4	118.3	146	328.5	584	912.5	1314	1788.5	2336	2956.5	3650
$\sqrt{1-.00000002990\,v_1^2\ell}$	.6	7¼	.814	3.26	7.32	13	20.4	29.3	39.9	52.1	65.9	81.4	183.2	325.6	508.8	732.6	917	1301.4	1647	2035
$\sqrt{1-.00000002673\,v_1^2\ell}$	.68	8¼	.483	1.93	4.35	7.73	12.1	17.4	23.7	30.9	39.1	48.3	108.7	193.2	301.9	434.7	591.7	772.8	978.1	1207
$\sqrt{1-.00000002075\,v_1^2\ell}$	.8	9⅝	.26	1.04	2.34	4.2	6.5	9.4	12.7	16.6	21.1	26	58.5	104	162.5	234	318.5	416	526.5	650
$\sqrt{1-.00000001816\,v_1^2\ell}$	.88	10⅝	.176	.704	1.58	2.82	4.4	6.34	8.62	11.26	14.26	17.6	39.6	70.4	110	158.4	215.6	281.6	356.4	440
$\sqrt{1-.00000001636\,v_1^2\ell}$	.96	11⅝	.1225	.49	1.1	1.96	3.06	4.41	6	7.84	9.92–9.93	12.25	27.6	49	76.6	110	150	196	248	306
$\sqrt{1-.00000001466\,v_1^2\ell}$	1.04	12½	.09	.36	.81	1.44	2.25	3.24	4.41	5.76	7.3	9	20.25	36	56.25	81	110.25	144	182.25	225
$\sqrt{1-.00000001362\,v_1^2\ell}$	1.12	13½	.067	.268	.603	1.07	1.68	2.41	3.28	4.29	5.43	6.7	15.1	26.8	41.9	60.3	82.1	107.2	135.7	167.5
	2	3	4	5	6	7	8	9	10	11	12	13	14	15	16	17	18	19	20	21

Remark — ℓ is the length of Main in Feet. V_1 is the Velocity of Air at Entrance to Main in Feet per Second.

Values of V_1^2

FIG. 5.—Friction Losses of Air in Pipes.

Ratio of — Absolute Pressure at Lower End of Main to Absolute Pressure at Entrance	Inside Diameter of Pipe		— Cubic Feet of Free Air Compressed per Minute to — 90 Lbs Gauge per Square Inch.																
	Feet	Inches	100	200	300	400	500	600	700	800	900	1000	1500	2000	2500	3000	3500	4000	4500	5000
$\sqrt{1-.00000046 v_1^2 \ell}$	.166	2	112.15	448.6	1009.4	1794.4	2803.8	4037.4	5495.4	7177.6	9084.15	11215	25233.5	44860						
$\sqrt{1-.00000177 v_1^2 \ell}$	.25	3	22.56	90.24	203.04	361	564	812.16	1105.4	1443.3	1827.0	2256	5076	9024	14100	20304	27636	36096	45664	56400
$\sqrt{1-.00000658 v_1^2 \ell}$	.33	4	7.13	28.52	64.2	114.1	178.25	256.7	349.4	456.5	577.5	713	1604.3	2852	4456.3	6417	8734.5	11408	14438.3	17825	
$\sqrt{1-.00000492 v_1^2 \ell}$	.415	5	2.92	11.68	26.3	46.7	73	105.4	143.1	186.9	236.5	292	657	1168	1825	2628	3577	4672	5913	7300
$\sqrt{1-.00000370 v_1^2 \ell}$	.52	6¼	1.2	4.8	10.8	19.2	30	43.2	58.8	76.8	97.2	120	270	480	750	1080	1470	1920	2430	3000
$\sqrt{1-.00000299 v_1^2 \ell}$	.6	7¼	.696	2.6	5.9	10.6	16.5	23.8	32.3	42.2	53.5	66.6	148.5	264	412.5	594	808.5	1056	1336.5	1650
$\sqrt{1-.00000257 v_1^2 \ell}$	.68	8¼	.39	1.56	3.6	6.24	9.75	14.04	19.1	25	31.6	39	87.8	156	243.8	351	477.8	624	789.8	975
$\sqrt{1-.00000207 v_1^2 \ell}$	.8	9⅝	.21	.84	1.9	3.36	5.26	7.56	10.3	13.44	17	21	47.3	84	131.3	189	257.3	336	425.3	525
$\sqrt{1-.00000183 v_1^2 \ell}$	.88	10⅝	.14	.56	1.26	2.24	3.5	5.04	6.9	8.96	11.3	14	31.5	56	87.5	126	171.5	224	283.5	350
$\sqrt{1-.00000163 v_1^2 \ell}$	.96	11⅝	.1	.4	.9	1.6	2.5	3.6	4.9	6.4	8.1	10	22.5	40	62.5	90	122.5	160	202.5	250
$\sqrt{1-.00000146 v_1^2 \ell}$	1.04	12½	.08	.32	.72	1.28	2	2.88	3.9	5.12	6.5	8	18	32	50	72	98	128	162	200
$\sqrt{1-.00000136 v_1^2 \ell}$	1.12	13½	.053	.212	.48	.85	1.325	1.91	2.6	3.41	4.4	5.3	11.9	21.2	33.13	47.7	64.9	84.8	127.6	132.5
			2	3	4	5	6	7	8	9	10	11	12	13	14	15	16	17	18	19	20	21

— Values of $\frac{v_1^2}{7}$ —

— Remark. — ℓ is the Length of Main in Feet. — v_1 is the Velocity of Air at Entrance to Main in Feet per Second. —

Fig. 6.—Friction Losses of Air in Pipes.

COMPRESSED AIR. 27

Ratio of Absolute Pressure at Lower End of Main to Absolute Pressure at Entrance	Inside Diameter of Pipe		Cubic Feet of Free Air Compressed per Minute to 100 Lbs Gauge per Square Inch.														
	Feet	Inches	100	200	300	400	500	600	700	800	900	1000	1500	2000	2500	3000	3500	4000	4500	6000
$\sqrt{1-.0000002045\, v_1^2 \ell}$	.166	2	34.4	137.6	309.6	550.4	860	1238.4	1685.6	2201.4	2786.9	3440	7740	13760	21600	30960	42140	55040	69660	86000
$\sqrt{1-.00000177\, v_1^2 \ell}$	.25	3	6.81	27.24	61.2	108.9	170.3	245.2	333.7	435.84	551.6	681	1532.3	2724	4256.3	6129	8342	10896	13790	17225
$\sqrt{1-.000000685\, v_1^2 \ell}$	.33	4	2.16	8.64	19.44	34.6	54	77.8	105.8	138.24	175	216	486	864	1350	1944	2646	3456	4374	5000
$\sqrt{1-.0000004472\, v_1^2 \ell}$	.415	5	.88	3.52	7.92	14.1	22	31.7	43.2	56.32	71.3	88	198	352	550	792	1078	1408	1782	2200
$\sqrt{1-.000000379\, v_1^2 \ell}$	.52	6¼	.361	1.44	3.24	5.8	9.03	13	17.7	23.1	29.2	36.1	81.2	144.4	225.6	324	441	576	729	900
$\sqrt{1-.000000299\, v_1^2 \ell}$	.6	7¼	.2	.8	1.8	3.2	5	7.2	9.8	12.8	16.2	20	45	80	126	180	245	320	405	500
$\sqrt{1-.000000253\, v_1^2 \ell}$	.68	8¼	.115	.46	1.04	1.84	2.88	4.14	5.64	7.36	9.3	11.5	25.9	46	71.9	103.6	140.9	184	233.9	287.5
$\sqrt{1-.0000002075\, v_1^2 \ell}$	.8	9⅝	.063	.25	.59	1	1.58	2.27	3.08	4.02	5.1	6.3	14.2	25.2	39.4	56.7	77.2	100.8	127.6	157.5
$\sqrt{1-.0000001835\, v_1^2 \ell}$	.88	10⅝	.043	.172	.39	.69	1.08	1.55	2.1	2.75	3.5	4.3	9.7	17.2	26.9	38.7	52.7	68.8	87.1	107.5
$\sqrt{1-.0000001636\, v_1^2 \ell}$	.96	11⅝	.03	.12	.27	.48	.75	1.08	1.47	1.92	2.4	3	6.75	12	18.8	27	36.75	48	60.75	75
$\sqrt{1-.0000001466\, v_1^2 \ell}$	1.04	12½	.023	.092	.21	.37	.58	.83	1.13	1.47	1.9	2.3	5.2	9.2	14.4	20.7	28.2	36.8	46.6	57.5
$\sqrt{1-.0000001362\, v_1^2 \ell}$	1.12	13½	.016	.064	.144	.26	.4	.58	.78	1.024	1.3	1.6	3.6	6.4	10	14.4	19.6	25.6	32.4	40
			2	3	4	5	6	7	8	9	10	12	13	15	16	17	18	19	20	21

— Values of v_1^2 —

Remark.— ℓ is the Length of Main in Feet. — v_1 is the velocity of Air at Entrance, in Feet per Second.

FIG. 7.—Friction Losses of Air in Pipes.

that the ratio of absolute pressures at lower and upper end of main is:

$$\sqrt{1 - 0.00000002075\ v_1{}^2\ l}$$

and as we know that this ratio is equal to 0.971, we may write:

$$0.971 = \sqrt{1 - 0.00000002075\ v_1{}^2\ l}$$

Or, squaring both members of this equation:
$$0.9428 = 1 - 0.00000002075\ v_1{}^2 \times 5280 \times 5$$
Or: $\quad 0.0005478\ v_1{}^2 = 1 - 0.9428$
hence: $\quad v_1{}^2 = 104.4$

which we must find in the horizontal column starting from 9-⅝″; we see that this number is comprised between 84 (2000 cu. ft.) and 131.3 (2500 cu. ft.).

The required number is intermediate between 2000 and 2500 cu. ft.; it can, with sufficient accuracy, be obtained by interpolation:

131.3—84=47.3. Corresponding to a difference of 500 cu. ft. of free air (from 2500 to 2000).

104.4—84=20.4, which, by a simple rule of three, corresponds to: $500 \times \frac{20.4}{47.3} = 215$, and the required number of cubic feet of free air per minute is:

2215.

EXAMPLE 3.

We desire to convey 1000 cu. ft. of free air per minute, compressed to 70 lbs. gauge, through a pipe 3 miles long, the loss in pressure not to exceed 5 lbs. What must be the diameter of the pipe?

This diameter could be determined directly, but through calculations more intricate than by the tables, which can be used in the following manner:

The pressure at entrance to main is   84.7 lbs. absolute.
The permissible loss is   5.
The pressure at the lower end of main is:
    79.7 lbs. absolute,
and the percentage of loss is:
$$\frac{79.7}{84.7} = 0.94.$$

Referring to Table Fig. 4 (70 lbs. gauge) the right value of $v_1{}^2$ is somewhere in the vertical column headed 1000.

The length is 15840 feet = l.

We will try some values of $v_1{}^2$ and apply them to the corresponding ratio of terminal pressures, until the result is exactly or approximately 0.94.

If the result is not exactly 0.94 we will then take the nearest larger commercial size of pipe, thus giving less than 5 lbs. loss through the main.

To facilitate these approximations we may remark that,

using the formula of Col. I, we will have an expression of this form:

$$0.94 = \sqrt{1 - 0.0000000 **** v_1^2 \, l}$$

in which the stars represent some numerical value to be discovered; or, squaring both members of this equation:

$$0.0000000 **** v_1^2 \, l = 1 - 0.94^2$$
$$= 0.1164$$

Let us try $v_1^2 = 449$, corresponding to a 5-in. pipe, we have
$$0.00000004972 \times 449 \times 15840 = 0.3536$$
which result is much too large.

We see that we have evidently to take a smaller value of v_1^2 since l remains constant, while the factor corresponding to 4972 diminishes with v_1^2.

Trying $v_1^2 = 100$, which corresponds to a 7¼-inch pipe, we find:
$$0.0000000299 \times 100 \times 15840 = 0.0474$$
which is below the value 0.1164 which we desire.

Taking $v_1^2 = 183$, which corresponds to a 6¼-inch pipe, we have:
$$0.00000003709 \times 183 \times 15840 = 0.1075.$$

This is the nearest value smaller than 0.1164 and will give less than 5 lbs. less; and thus we conclude that the required diameter of pipe is 6¼ ins.

A short use of the tables will render them quite convenient to use:

The above three examples cover the principal question liable to arise in ordinary practise, and the few calculations involved are more than balanced by the greater correctness of the results derived from Unwin's formulæ.

We can use the tables to find the loss of pressure incurred in the passage of air through a pipe of a given diameter and length, and with a given velocity of ingress. But it is interesting to know at the same time the corresponding loss of power.

With this object in view, a Table (Fig. 9) and curves (Fig. 8) are here given, showing the ratio of available power at full expansion and without reheating at the lower end of the main to the available power at full expansion and without reheating at its entrance.

These curves show that the comparative loss of power is always smaller than the comparative loss of pressure, and they will be found useful in estimating the total loss incurred in a given transmission.

Each curve corresponds to a certain pressure at the entrance to the main, these pressures being, as above, 70, 80, 90, and 100 lbs. gauge.

This addition to the study of the frictional losses is intended to dispel the confusion frequently made between the loss of pressure and the loss of power, there being a common tendency to consider those two terms as equivalent.

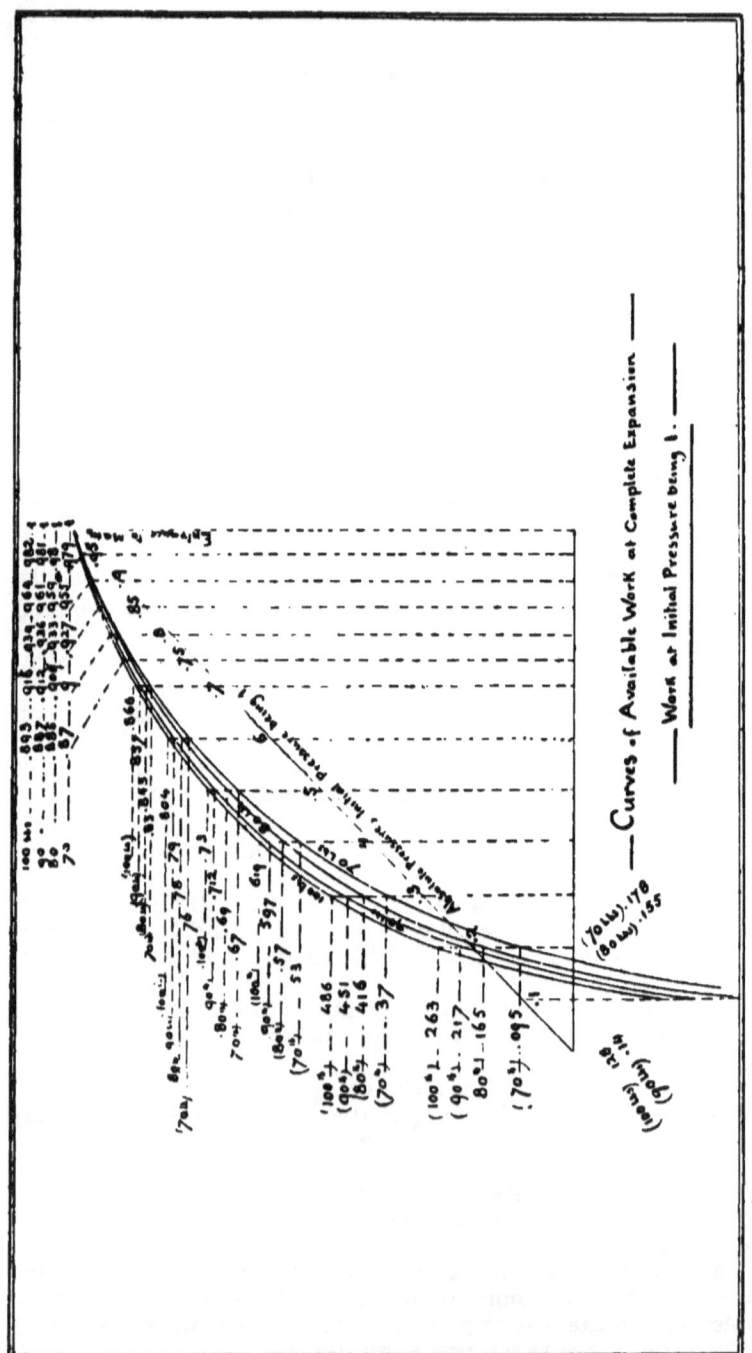

Fig. 8.

Absolute Pressure, the Absolute Press.re at Entrance to Main being taken as Unit	Available Work at Full Expansion the Work at Entrance Press.re (70 lbs g) being taken as Unit	Available Work at Full Expansion the Work at Entrance Press.re (80 lbs g) being taken as Unit	Available Work at Full Expansion the Work at Entrance Press.re (90 lbs g) being taken as Unit	Available Work at Full Expansion the Work at Entrance Press.re (100 lbs g) being taken as Unit.
1.00	1.000	1.000	1.000	1.000
.95	.979	.9809	.9816	.982
.90	.955	.959	.961	.964
.85	.927	.933	.936	.939
.80	.900	.909	.912	.916
.75	.870	.885	.887	.893
.70	.830	.847	.857	.866
.60	.760	.780	.790	.804
.50	.670	.690	.712	.730
.40	.530	.570	.597	.619
.30	.370	.416	.451	.486
.20	.095	.165	.217	.263
.10	− 430	− .318	− .239	− .165

—— Comparative Values of Work at Full Expansion and Abs. Pressure ——
—— at Various points of an Air Main ——
—— for 70 – 80 – 90 – 100 Lbs Effective per Square Inch at Entrance ——

FIG. 9.

For instance, if air enters a pipe at 100 lbs. gauge pressure =114.7 absolute, and is discharged at 80 lbs. gauge, or 94.7 lbs. absolute, thus showing a reduction of 20 per cent of gauge pressure, it is popularly and erroneously estimated that the air has lost 20 per cent of its power. The ratio between the absolute pressures is $94.7/114.7 = 82.5$ per cent.

Referring to the 100 lbs. curve, we note that the ratio of pressures 0.825 lies midway between 92 and 94 per cent ratios of power; i. e., corresponds to 93 per cent, showing a loss of 7 per cent only, instead of the so-called loss of 20 per cent of power.

The explanation of this is, that, while the pressure is diminished, the volume is proportionately increased, and the real loss of power is the work which the air could perform in expanding isothermally from the higher pressure to the lower pressure, which work has been absorbed by friction.

An inspection of the curves (Fig. 8) will show that the actual pressure at each point on the main becomes a constantly decreasing fraction of the initial pressure as the distance of this point from the entrance becomes greater, and these variations of pressure are figured by a line at 45 degrees on the co-ordinate axes. For each absolute pressure, the value of the available power at full expansion, as compared with the available power at entrance, is carried on the corresponding ordinate, and by joining the ends of these ordinates, four curves of powers have been drawn, each one corresponding to one of the above-named initial pressures. Although a limited range of initial pressures has been considered, the following general deductions are suggested by an inspection of these curves:

1. So long as the fall of pressure remains below a certain value, which, in the cases considered, is about 50 per cent, the loss of pressure is more rapid than the loss of power, and the ratio of powers is greater than the corresponding ratio of pressures.

2. When the pressure continues to fall beyond this value, the loss of power becomes more rapid than the loss of pressures, the ratio of powers remaining, however, greater than the corresponding ratio of pressure.

3. When the absolute pressure in the main becomes—in the cases considered—from 15 to 25 per cent of the initial absolute pressure, the ratio of the powers becomes equal to the ratio of the pressures.

4. The pressure continuing to fall, the loss of power becomes much more rapid than the loss of pressure, until the pressure is equal to the atmosphere, when the available power naturally becomes 0, and then negative (a case which is not to be considered here), and the ratio of powers is smaller than the corresponding ratio of pressures.

The inspection of the curves also shows that the relative deficiency of the power, as compared with the corresponding pressure, occurs more rapidly with a low initial pressure than with a higher one; and incidentally confirms the foregone statement, that, for a given initial pressure and velocity at entrance, there is a limit of length to each particular size of main, beyond which neither pressure nor power would be obtainable at its lower end, the whole pressure having been absorbed in overcoming the friction, and the air issuing from the pipe at atmospheric pressure.

And as, on the other hand, the velocity of the air varies inversely with its pressure at entrance, the desirability of high pressure is apparent, either as permitting the use of a smaller pipe to convey a given weight of air, or as increasing the distance at which a certain power can be obtained with a given size of pipe. This statement refers to the conveyance of the air, and is, of course, irrespective of the convenience of producing a high air pressure.

LOSS OF PRESSURE IN THE PASSAGE OF AIR THROUGH BENDS.

In addition to the frictional loss incurred in the passage of air through a straight pipe, under given conditions of length, diameter, and velocity, another cause of resistance is due to the changes of direction in the flow of air.

The bends in a pipe line should be as few as possible, but whenever they are absolutely necessary, as, for instance, when leaving the surface of the ground to penetrate in a vertical or inclined shaft, abrupt bends should be avoided. The branching at right angles by means of a T, so frequently found in small-sized steam, air, or water pipe, should be absolutely discarded.

Iron pipes of small size can easily be bent to a larger radius, and as to larger pipes, special elbows should be used instead of the common fittings, whose radius is always small as compared with the diameter of the pipe.

The annexed table shows that when the mean radius of curvature is equal to the diameter of the pipe, the loss incurred in the air pressure is nearly four times as great as when the radius of curvature is equal to five diameters, so that a pipe line may be established with all possible care regarding its diameter and the velocity of the air, consistent with a small frictional loss, and much of the benefit derived therefrom be counteracted by the use of one or two short bends.

With reference to the table, it will be noticed that it applies to bends at a right angle. When a smaller or larger arc than 90 degrees is used, a sufficient approximation will be obtained in figuring the frictional loss in proportion to the length of arc of the bend, as compared with an arc of 90 degrees, and of same radius.

If possible, from 15 to 20 feet per second should be the average entrance velocity given to air in pipes less than 12 inches in diameter. Above this the velocity may be increased, but never to exceed 50 feet per second, for economical use.

Mean Radius of Curvature of Bend, expressed in Internal Diam. of Pipe as Unit	1	2	3	4	5
Loss in Lbs per Sq. Inch	$.005\, v_1^2$	$.0022\, v_1^2$	$.0016\, v_1^2$	$.0013\, v_1^2$	$.0012\, v_1^2$

Loss of Pressure in Lbs per Square Inch through 90° Bends

Remark.— v_1 is the velocity of Air in Feet per Second at Entrance.

Fig. 10.

THE INFLUENCE OF THE DIFFERENCE OF LEVEL ON THE USE OF COMPRESSED AIR.

The calculations concerning the applications of compressed air are generally based upon the standard values of the atmospheric pressure at the sea level; viz., 14.7 lbs. per square inch. The fact that a large number of mines are located at a considerable altitude makes it necessary to investigate the influence of this condition upon the use of compressed air, and it will be shown herein that the differences of level are not to be overlooked in designing a system for power transmission. The weight of one cubic foot of air, at the surface of the earth, and at 32 degrees Fahr., and when the barometer stands at 30 inches, is 0.0807 lbs. The position of the mercury in a barometer is due to the weight of a column of air. whose height would be the thickness of the atmospheric layer that surrounds the earth, and as one cubic inch of mercury weighs 0.491 lbs., the weight of a column of mercury 1 inch square and 30 inches high is $30 \times 0.491 = 14.73$ lbs. Hence the conclusion that a column of air 1 inch square and of the height of the atmosphere weighs 14.7 lbs., and will balance the weight of a column of mercury 1 inch square and 30 inches high.

The immediate consequence of this is that as we rise above the level of the sea at a given place, the atmospheric pressure per square inch must decrease, since the height of the column of atmosphere pressing on the mercury of the barometer diminishes, and we can readily calculate that if the whole atmospheric layer were of equal density, that is, if one cubic foot of air had the same weight at any altitude, the thickness of our atmosphere would be 26,208 feet, or 4 97 miles.

Such, however, is not the case. The weight of one cubic foot of air varies with its pressure and with its temperature, which both change with the altitude.

It is commonly assumed, that at the same latitude, the temperature drops by 1 degree Fahr. for every 340 feet of height above the sea level; but this could not be taken as anything like a general rule, since the temperature is affected by many local and variable conditions. It suffices, however, to show that the density of air changes with the altitude, but as the laws of this variation are imperfectly known, and only for moderate altitudes, the exact thickness of the *atmospheric layer that surrounds our planet is a matter of speculation. It is generally conceded, however, to be about 45 miles.

The variations of atmospheric pressure with the altitude have been, in the annexed table, calculated from the sea level to 10,000 feet above it, and for equal steps of 500 feet, on the assumption of a constant temperature of 60 degrees Fahr. prevailing throughout the change of altitude. This supposition, however, as we have mentioned before, is not correct, but the exact influence of the temperature can easily be computed for any particular instance.

An inspection of the table of atmospheric pressures leads to an immediate practical conclusion. Let us take, for instance, a machine designed to compress at the sea level 500 cubic feet of free air per minute to 80 lbs. gauge, that is, 80 lbs. above the atmospheric pressure. The volume of cold compressed air delivered per minute is

$$500 \times \tfrac{14.7}{94.7} = 77.6 \text{ cu. ft.}$$

Suppose now that the same compressor be used at 5000 feet altitude and run at the same number of revolutions; the piston will sweep through 500 cubic feet as before, but the atmospheric pressure being only 12.14 lbs. per square inch, the volume of cold air at 80 lbs. gauge delivered per minute will be

$$500 \times \tfrac{12.14}{92.14} = 65.85 \text{ cu. ft.}$$

That is to say, the delivery of air at 80 lbs. gauge and at 5000 feet altitude will be 85 per cent of the delivery at 80 lbs. gauge and at the sea level, from the same sized compressor running at the same number of revolutions.

These volumetric variations, reckoned upon the volume at the sea level taken as a unit, will be found recorded in four columns corresponding respectively to 70, 80, 90, and 100 lbs. gauge and annexed to the pressure column. It will be noticed that the volumetric efficiency, that is the ratio of the delivery at any given altitude to the delivery at the same pressure and at the sea level, decreases as the receiver pressure increases.

We know that in adiabatic compression (which we may take as a standard of comparison) the compression to 80 lbs. gauge and delivery of 500 cu. ft. of free air per minute absorbs 79.4 I. H. P. It may easily be calculated that for the same outside temperature (60 degrees Fahr.) and the same gauge pressure (80 lbs.) the compression and delivery at 5000 feet altitude of the same amount of atmospheric air will absorb 73.7 I. H. P.

The ratio of these powers is $\tfrac{73.7}{79.4} = .928.$

That is to say, we lose in capacity 15 per cent and we gain in power 7.2 per cent, which amounts to saying that the production at the same volume of air at the same effective pressure will require:

    1   I. H. P. at the sea level,
    1.093 "   at 5000 feet,
    1.190 "   at 10,000 feet altitude.

It costs more, therefore, to obtain the same useful work from a given compressor at high altitudes than at the sea level.

Four columns of I. H. P., referring to the compression of 100 cubic feet of free air per minute to 70, 80, 90, and 100 lbs. respectively, are recorded alongside of the volumetric results. An inspection of the table shows that if we compare the work absorbed by 1 cu. ft. of air delivered at a given pressure, at 10,000 feet altitude for instance, and at the sea level, the ratio will be practically the same within the whole range of pressures considered.

Altitude above Sea Level — Feet —	Absolute Pressure per Square Inch — Lbs —	Receiver Gauge Pressure (Lbs per Square Inch)							
		70		80		90		100	
		V	IHP	V	IHP	V	IHP	V	IHP
0	14.70	1.000	14.67	1.000	15.72	1.000	16.88	1.000	17.83
500	14.42	.986	14.60	.984	15.67	.983	16.81	.982	17.76
1000	14.15	.971	14.49	.969	15.57	.967	16.65	.966	17.62
1500	13.88	.958	14.39	.953	15.44	.952	16.62	.951	17.47
2000	13.61	.942	14.28	.936	15.31	.935	16.37	.934	17.24
2500	13.35	.925	14.16	.922	16.04	.919	16.23	.918	17.18
3000	13.10	.910	14.06	.907	16.04	.904	16.10	.903	17.02
3500	12.85	.895	13.93	.890	14.93	.889	15.97	.888	16.86
4000	12.61	.878	13.84	.876	14.82	.874	15.82	.872	16.76
4500	12.37	.867	13.71	.864	14.70	.860	15.70	.859	16.60
5000	12.14	.853	13.60	.849	14.58	.846	15.58	.844	16.44
5500	11.91	.838	13.52	.835	14.42	.832	15.42	.830	16.31
6000	11.68	.826	13.39	.822	14.32	.818	15.29	.816	16.15
6500	11.46	.813	13.28	.808	14.19	.804	15.15	.802	16
7000	11.24	.798	13.16	.793	14.07	.791	15.02	.788	15.87
7500	11.03	.786	13.06	.780	13.94	.777	14.91	.776	15.78
8000	10.82	.773	12.97	.767	13.84	.764	14.77	.761	15.69
8500	10.62	.762	12.85	.754	13.72	.751	14.63	.749	15.42
9000	10.42	.746	12.73	.742	13.54	.739	14.52	.736	15.29
9500	10.22	.734	12.63	.729	13.46	.726	14.40	.723	15.14
10000	10.03	.722	12.55	.716	13.38	.714	14.28	.711	15.05

— Variations —
— of Volumetric Efficiency and of Work of Compression —
— at various Altitudes and Receiver Pressures. (Temperature 60°Fahr.) —

— Remarks.— Letter V means: Relative Volume of Air delivered at each Receiver Pressure.
— Letters I.HP mean: Indicated Horse-Power per 100 cubic feet of Free Air and per Minute. —

Fig. II.

This is not the only effect of a difference of altitude and a practical case will illustrate another side of the question:

Suppose that a mining plant is located 1500 feet above the Compressor plant, and that the Compressor plant itself is situated at an altitude of 3000 feet above the sea level, and that the receiver pressure at the compressor is 80 lbs. The atmospheric pressure at the elevation of 3000 feet in the Compressor room is 13.1 lbs. per square inch. One cubic foot of air at the sea level, and at 60 degrees Fahr. weighs 0.0764 lbs. One cubic foot of air at 3000 feet elevation and 60 degrees Fahr. will weigh

$$0.0764 \times \tfrac{13.1}{14.7} = 0.0681 \text{ lbs.}$$

Or, 1 lb. of air will represent a volume of 14.68 cubic feet. This volume represents a vertical column one inch square and 2113.92 feet high at the pressure of 13.1 lbs. per square inch, and at a pressure of 80 lbs. gauge or 93.1 lbs. absolute, the height of this column weighing 1 lb., and 1 inch square in section is

$$2113.92 \times \tfrac{13.1}{93.1} = 298.06 \text{ ft.}$$

Consequently a column of air at 80 lbs. pressure, 1500 feet high, represents a pressure of 5.03 lbs. per square inch.

The absolute pressure of air, which at the lower end of the pipe is 93.1 lbs., is at the upper end:

$$93.1 - 5.03 = 88.07 \text{ absolute.}$$

and as the atmospheric pressure at 4500 ft. is 12.37 lbs. the effective pressure at the hoisting works is 88.07—12.37 lbs., or 75.7 lbs. So there is, regardless of the loss due to friction in this respect, no loss of volume, but a loss of pressure.

A very similar course of reasoning would show that when compressed air is carried down a shaft the pressure at the lower end is greater than the receiver pressure, and this excess of pressure, due to the weight of this column of air, will generally more than balance any frictional losses there may be in the pipes.

It must be remembered, in this connection, that any motors operated by this compressed air will also have a larger back pressure to encounter in the exhaust than they would at the mouth of the shaft, but still the loss due to this back pressure is only a small portion of the gain by the difference of level.

In both of these examples an exact computation would require a consideration of the temperature, but which may be neglected in all ordinary propositions.

AIR ENGINES.

Compressed Air, like all elastic gases, can be made to operate a piston by its expansive force, exactly as does steam, and it may be stated in a general way, that any steam engine can be actuated by air without altering its arrangement. It is, moreover, hardly necessary to add that this statement applies to the non-condensing steam engines only.

Tables are herewith given of the consumption of air per minute, reduced to atmospheric pressure, in three classes of engines more commonly used; viz:

The Slide Valve Engine,
The Automatic Cut-off Engine, Single and Compound,
The Corliss Engine, Single and Compound.

Air and Steam, however, while partaking of the same general active property, differ widely in several respects, and a few explanatory remarks are here necessary.

In the first place, the pressure of air may be independent of its temperature. This valuable feature, which makes otherwise the use of compressed air so convenient, is fraught, however, with practical consequences which in many cases, and unless provided for, would render it impossible.

Air, in most cases, expands in a motor adiabatically; i. e., its expansion is accompanied by a considerable fall of temperature. An additional table is here presented (Fig. 19), giving the temperature of exhaust of air, after working expansively in the various types of engines considered. This temperature is found to range from $+7.5$ in the slide valve engine to -143 in the Compound Corliss, cutting off at $\frac{1}{3}$ of stroke, the air being admitted to the engine at 60 degrees Fahr., and while the former temperature might not prove troublesome with dry air, on account of the strong exhaust blast of an engine with a late cut-off, the latter is decidedly unacceptable, as any lubricant introduced in the cylinder would freeze instantly, and the exhaust ports be promptly clogged with ice, especially in the interior of a mine where the moisture of air is more marked than outside.

It will therefore be necessary for the economical use of air to heat it to a certain extent, either before it enters the motor, or during the process of its expansion within the cylinder. We know already that this operation has also the effect of increasing the volume of the air at constant pressure. Two curves are here presented showing the increase of volume of 1 cubic foot of air, at 32 degrees Fahr. and at 60 degrees Fahr., when heated to various temperatures up to 500 degrees Fahr.

In connection with this subject of re-heating, another distinctive feature of air as compared with steam must be pointed out.

In all non-condensing steam engines, even with an early cut-off, the proportions are such as to maintain at the end of the period of expansion, a sufficient steam pressure to insure a speedy exhaust of the gaseous and of the condensed steam. This pressure must of course be greater in a fast moving than in a slow engine, with the consequence that part of the energy of the steam is thus sacrificed, not uselessly, indeed, but without doing useful work.

But with air, there is no condensation during the expansion, and also the active gas which operates the piston being the

Cubic Feet of Free Air at 14.7 Lbs. per Sq. Inch (Absolute) consumed per Minute in Slide Valve Engines

Size of Engine Inches	Revolutions per Minute	Piston Velocity per Minute Feet	60 lbs		70 lbs		80 lbs	
			cu ft	BHP	cu ft	BHP	cu ft	BHP
6 × 8	250	333	232.7	8.6	264	10	295	11.5
7 × 10	240	400	380	14	431	16.4	482	18.8
8 × 10	240	400	496.3	18.3	562.8	21.5	620.2	24.5
9 × 12	200	400	628.4	23.2	712.5	27.2	796.6	31
10 × 12	200	400	775.7	28.7	879.6	33.6	988.4	38.4
10 × 14	200	467	904.8	33.5	1025.9	39.2	1147	44.8
11 × 14	200	467	1094.7	40.6	1241.3	47.5	1387.8	54.1
12 × 16	180	480	1341.1	49.7	1520.6	58	1699.2	66.5

Remarks — Clearance is assumed to be 7% of Cylinder Capacity. — Cut-off at ⅝ of Stroke. — Brake Horse Power is taken as .85 of Indicated Horse Power. — Initial Pressure in Cylinder is taken as .95 of Pressure at Throttle. —

FIG. 12.

Size of Engine Inches	Revolutions per Minute	Piston Velocity per Minute Feet	60 Lbs		70 Lbs		80 Lbs		90 Lbs	
				BHP		BHP		BHP		BHP
7 × 9	300	450	cu.ft 162.9	12.8	cu.ft 207.4	15.8	cu.ft 251.8	17.8	cu.ft 255.6	20.4
	360	540	219.5	15.8	248.8	18.4	278.2	21.4	306.7	24.4
8 × 9	300	450	259.8	16.7	271.9	20	304	23.8	335.1	26.6
	360	540	288	20	326.6	24	365.1	28	402.6	32
9 × 9	300	450	303.3	23.1	342.7	26.4	383.2	24.6	422.5	30.7
	360	540	362.7	25.3	411.3	30.4	459.8	35.4	506.9	40.4
9½ × 10½	270	472.5	353	23.9	400.8	28.7	447.6	33.4	493.5	38.1
	330	577.5	431.2	28.7	489	34.4	546.8	40	602.8	45.7
10½ × 10½	270	472.5	402.2	29.2	440.2	36.1	548	40.8	604.2	46.6
	330	577.5	527.8	35	598.6	42.1	669.1	44	727.7	56
11 × 12	240	480	482.6	32.5	547.2	39.1	611.8	46.8	674.5	62
	300	600	603.5	40.6	624.3	40.9	746.1	56.8	848.6	64.9
12½ × 12	240	480	623.2	42	706.8	60.5	790.2	51.7	871.1	67.1
	300	600	778.8	52.5	883	63.1	917.8	73.4	1022.4	83.9

— Gauge Pressure at Throttle in Lbs per Square Inch —
— and Corresponding Brake Horse Power —

— Cubic Feet of Cold Air at 14.7 Lbs Absolute per Square Inch —
— consumed per Minute in —
— Single-Cylinder Automatic Cut-Off Engines. —

Remarks.— Clearance is assumed to be 5% of Cylinder Capacity.
— Cut-off at ¼ of Stroke.
— Brake Horse-Power is taken as .85 of Indicated Horse-Power.
— Initial Pressure in Cylinder is taken as .96 of Pressure at Throttle.

FIG. 13.

COMPRESSED AIR.

Gauge Pressure at Throttle in Lbs per Square Inch and corresponding Brake Horse-Power

Size of Engine Inches	Revolutions per Minute	Piston Velocity per Minute Feet	90 Lbs	BHP	100 Lbs	BHP
			cu.ft		cu.ft	
9½ and 14½ by 10½	260	455	778.9	62.2	855.7	70.66
" "	290	507.5	869	69.4	954.7	78.6
10½ and 16 by 12	245	490	1024.5	81.3	1125.5	92.3
" "	275	550	1150.2	91.4	1263.6	103.7
12 and 17½ by 13¾	230	517.5	1414.3	113	1553.8	128.9
" "	260	585	1598.5	126.6	1756.1	144

Cubic Feet of Cold Air at 14.7 Lbs Absolute per Sq. Inch consumed per Minute in Compound Automatic Cut-off Engines

Remarks.— Clearance is assumed to be 5% of Cylinder Capacity.— Cut-off at 7/16 of Stroke in H.P. Cylinder. Brake Horse Power is taken as .85 of Indicated Horse-Power. Initial Pressure in H.P. Cylinder is taken as .95 of Pressure at Throttle.

FIG. 14.

COMPRESSED AIR.

Size														
10 × 24	90	360	353.6	27	290.4	22	389.8	31	320.2	26	428.2	36	351.8	30
10 × 30	90	450	441.8	33	363.2	28	487	39	400.4	33	535.1	44	440	37
12 × 30	90	450	637.6	48	523.6	40	702.9	56	577.2	47	772.2	64	634.1	54
12 × 36	90	540	764.4	58	627.9	48	842.8	67	692.2	56	925.9	77	760.5	65
13 × 30	90	450	746.4	56	613	47	822.9	66	675.9	55	904	75	742.6	63
13 × 36	90	540	895.8	68	736	56	987.6	79	811.5	66	1085	90	891.5	76
14 × 36	90	540	1042	79	855.9	66	1148.8	92	943.6	77	1262	106	1036.6	88
14 × 42	85	595	1148.3	87	942.8	73	1265.9	101	1039.4	85	1390.7	117	1141.9	97

— Cubic Feet of Free Air at 14.7 Lbs Absolute per Square Inch Consumed per Minute in Single-Cylinder Corliss Engines. —

Remarks: — Clearance is assumed to be 3% of Cylinder Capacity. — Brake Horse-Power is taken as .85 of Indicated Horse-Power — Initial Pressure in Cylinder is taken as .95 of Pressure at Throttle. —

Size of Engine	Revolutions per Minute	Piston Velocity per Minute Feet	Effective Pressure at Throttle — 100 Lbs per Square Inch	Brake H.P.
Inches			Cu. Ft	
10 and 15 by 30	85	425	651	53
12 and 18 by 36	83	498	1099	94
14 and 22 by 36	83	498	1496	126
14 and 22 by 42	75	525	1577	133

— Cubic Feet of Free Air at 14.7 Lbs per Sq. Inch (Absolute) —
— consumed per Minute in —
— Compound Corliss Engines —

— Remarks.— Clearance is assumed to be 3% of H.P. Cylinder Capacity.
— Cut-off at ¼ of Stroke in H.P. Cylinder.
— Brake Horse Power is taken as .85 of Indicated Horse-Power.
— Initial Pressure in H.P. Cylinder is taken as .95 of Pressure at Throttle.

FIG. 16.

COMPRESSED AIR. 45

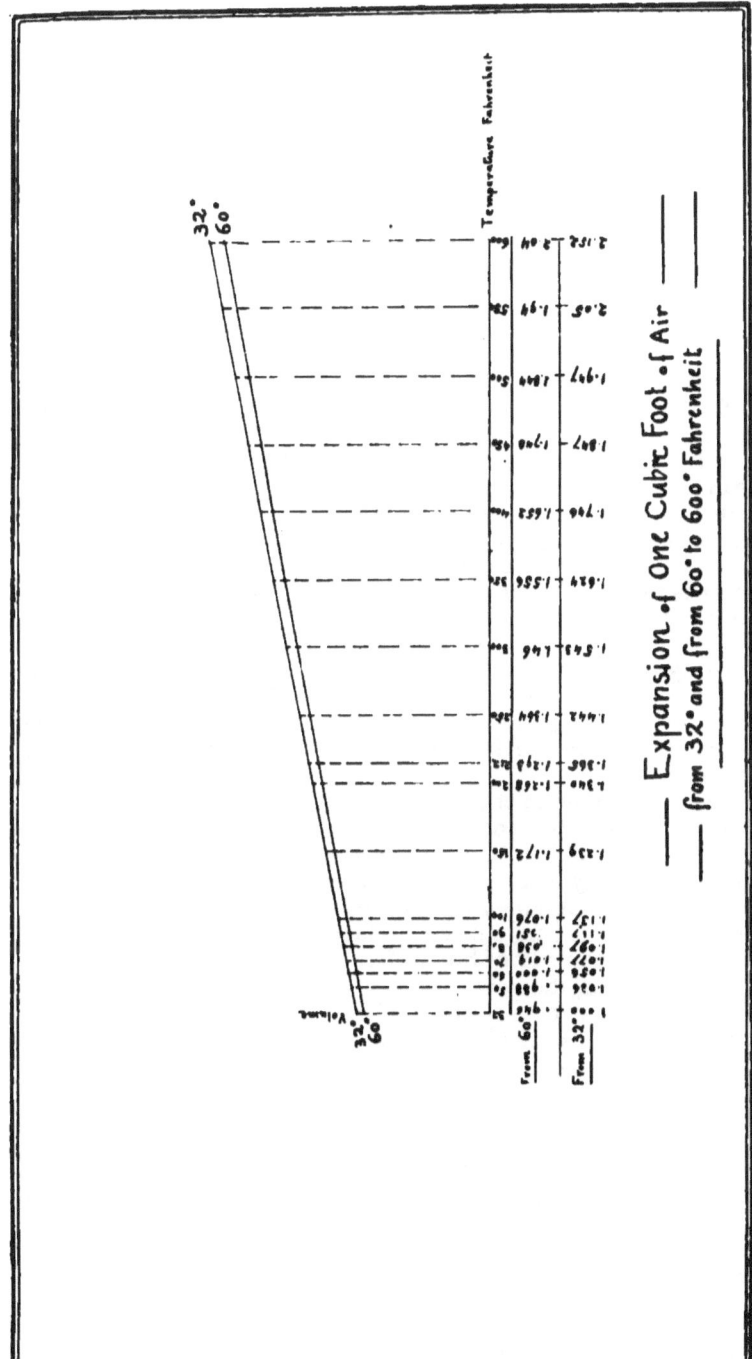

Fig. 17.

same as the medium into which it is discharged, the exhaust pressure may become a very insignificant quantity.

The result of this is two-fold: First, an air motor, unlike a steam engine, can work practically at complete expansion, i. e., the compressed air can expand into the cylinder down to atmospheric pressure; and second, this more prolonged expansion will be accompanied by a greater fall of temperature. So that it may be said that the genuine air motor is inseparable from a system of reheating, and also that the complete expansion of the air producing a greater variation of load on the piston, and of strains on the pieces, an air motor should not necessarily but preferably be a compound, rather than a single machine. For similar reasons it may rationally be expected that turbo-motors of the Parsons' type and DeLaval Rotary Engines would be especially well adapted to show a high efficiency as air motors.

It may be inferred, that while an ordinary steam engine will perform satisfactory duty if operated with air, a less consumption of it will be obtained by cutting off earlier in the stroke so as to work at complete expansion. This will diminish the mean effective pressure throughout the stroke, and, consequently, the power developed by the engine, at the same time extending the range of variation of the strains.

Such a state of affairs may be acceptable if the load on the motor is regular, but if—as will often be the case, especially in mining machinery—the load constantly varies or else is intermittent, the air motor at complete expansion must have its valve gear so arranged as to permit a later cut-off and, of course, a greater or smaller amount of exhaust pressure, which amounts to saying that it must be an ordinary steam engine susceptible of an earlier cut-off than is commonly used with steam.

Reverting now to the subject of reheating, several systems have been suggested and used.

If the only object was to preclude the obstruction of the exhaust ports by the formation of ice due to the moisture of air, it would be obtained by the application to this portion of the engine, of some source of heat, such as a lamp, or an injection of steam, or of hot water.

This process, however, hardly deserves more than a mere mention, for if such a source of heat is handy, it can be used to far better advantage in heating the air, either in the cylinder or before entering it.

One method consists in injecting into the cylinder a spray of warm water, whose heat is absorbed by the air, while the water is cooled. The annexed table gives the weight of water at 75 degrees, 100 degrees, and 150 degrees Fahr. to be supplied for each pound of air expanding to the atmospheric pressure from 70, 80, 90, and 100 lbs. gauge, so that the final temperature of air will be 32 degrees Fahr., its initial temperature being 60 degrees Fahr.

Gauge pressure of air.	B. T. U. required per lb. of air.	Pounds of water per lb. of air, the temperature of water being		
		75° Fahr.	100° Fahr.	150° Fahr.
lbs.	(A)	lbs.	lbs.	lbs.
70	59.	1.37	.86	.5
80	62.8	1.46	.92	.53
90	66.2	1.54	.97	.56
100	69.2	1.61	1.02	.6

Another and better method is to inject steam instead of hot water into the cylinder. The advantages of this system are, first that steam, being in a gaseous state, mixes up with air more readily than water, even finely pulverized, and besides, the condensation of this steam gives up its latent heat, which increases considerably the heating of air.

A comparison of this process with the previous one can readily be made. Assuming that a spray of water at 212 degrees Fahr. is injected into the cylinder, each pound of this water will give up 180 B. T. U. before it is cooled to 32 degrees Fahr.

But, taking steam at atmospheric pressure, i. e., also at 212 degrees Fahr., 1 lb. of steam, in the process of liquefaction, will abandon 966 B. T. U., its latent heat of vaporization, and besides 180 B. T. U. as above, making a total of 1146 B. T. U.

The following table gives the weight of steam at 212 degrees Fahr. required for each pound of air to prevent its temperature from falling below 32 degrees Fahr. at complete expansion.

Gauge pressure of air. Lbs.	B. T. U. required for each lb. of air.	Lbs. of steam at 212 degrees per lb. of air.
70	59.0	.051
80	62.8	.055
90	66.2	.059
100	69.2	.0604

It is evident that quite similar calculations could be made to maintain the exhaust temperature at any given point. Besides, the use of steam keeps the walls of the cylinder wet, and while water alone is a poor lubricant between metallic surfaces, it facilitates the action of the regular lubricants, and is also favorable to the tightness of the piston packing.

It will readily be seen that both these methods completely preclude the formation of ice in the exhaust ports; their good effect is still more pronounced if the cylinder is provided with a jacket, into which hot air is circulated.

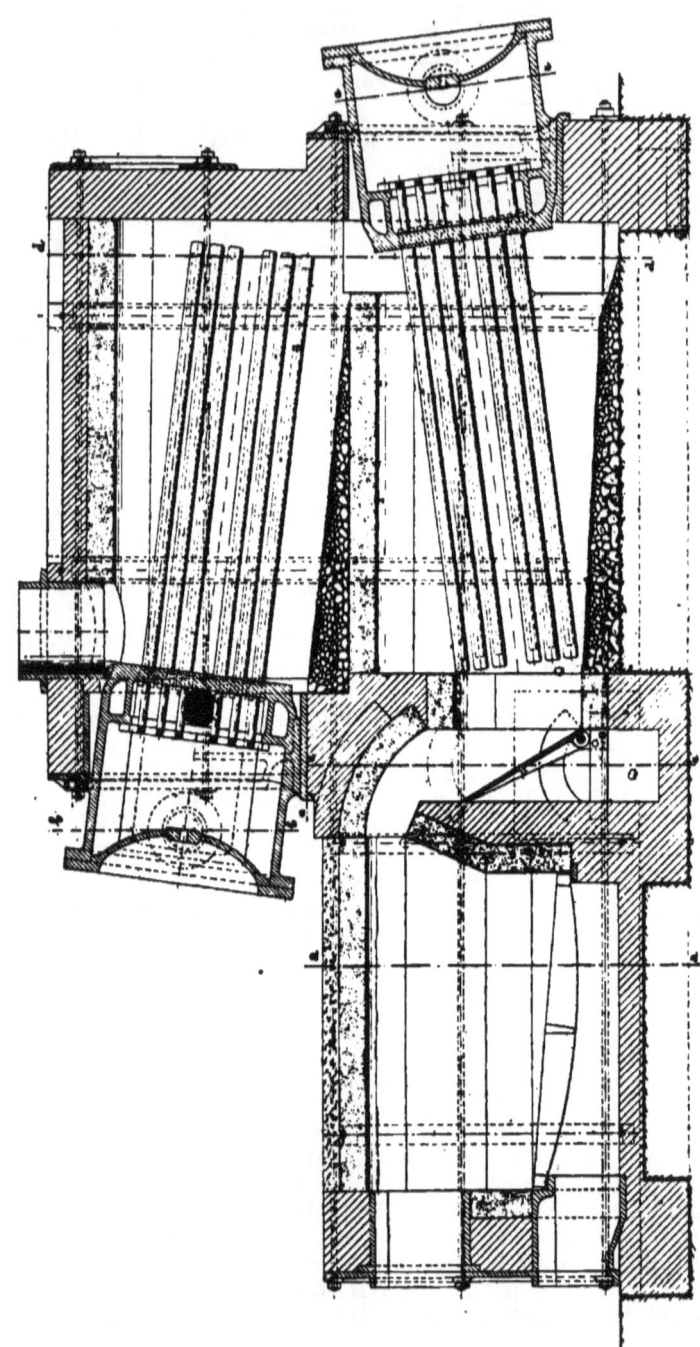

FIG. 18.—Rix Compound Pneumatic Reheater. Manufactured by Fulton Engineering and Shipbuilding Works, S. F., Cal.

Air can also be reheated before being admitted into the cylinder. Various designs of heaters are used for this purpose, the air generally passing through a system of pipes heated by an interior furnace, a flue being provided for the passage of the hot gases on the outside of the pipes before they reach the chimney. And as air, on account of its bad conductivity, does not easily take up heat from the metallic sides of the pipes, it is expedient to inject in the pipes a small quantity of water which absorbs the heat more readily and penetrates with the hot air into the cylinder.

Another method of heating is to place a lamp or gas jet within the air pipe. The use of coal or wood is not advisable in this case, as grit and cinders would be carried by the current of air into the motor.

Reheating by the electric current is still in the experimental state.

When the motor is a compound machine, the air should be again reheated after it has done work in the H. P. cylinder, and before it is admitted to the L. P. cylinder.

The table giving the temperatures of exhaust in cold air work, also gives the temperatures at which air should be reheated prior to its admission to the single engines, or to each cylinder of the compound engines, in order to exhaust at 32 degrees Fahr.

These temperatures are moderate, and can be obtained with hot water, or low pressure steam. If the heating is done by passing the air through heated pipes, the fuel consumption will be very small, as practise shows that 1 lb. of coal gives the air from 8,000 to 10,000 B. T. U. in a properly designed heater.

To utilize the full benefit of reheating, and of air expansion in compound engines, an early cut-off is very desirable. This can be accomplished by reheating to 350 degrees before the air enters each cylinder, and Table Fig. 20 shows the amount of free air required for various horse powers under this condition. A comparision with Table Fig. 16 will show the marked advantage of this arrangement.

Fig. 20½ shows a compound direct connected Corliss Hoisting Engine, built by the Fulton Engineering and Shipbuilding Company, in conformity with the data in Fig. 20. The air is twice reheated; that is to say before entering the high pressure cylinder, and also before entering the low pressure cylinder.

For convenience in estimating the power required to compress air and the amount of air which will be furnished by given powers, the following tables have been constructed:

The table in Fig. 21 shows the amount of cubic feet of free air at 60 degrees Fahr. and 14.7 lbs. absolute pressure per square inch, that can be compressed and delivered per minute per I. H. P., adiabatically, in a single stage jacketted cylinder compressor, in a two-stage compound jacketted compression and in isothermal compression.

Terminal Temperatures with Cold Air, and Amount of Reheating. — Air Engines

Class of Engine	Size Inches	Temperature Fahrenheit of Air — Entrance			Temperature Fahrenheit of Air — Exhaust			Clearance compared to capacity of Single or H.P. Cylinder	Cut-off in Fraction of Stroke	Temperature Fahrenheit of Reheating to exhaust at 32°Fahrenheit in each Cylinder		
		Single	Compound H.P.	Compound L.P.	Single	Compound H.P.	Compound L.P.			Single	H.P.	L.P.
Slide Valve		60			+7.5			.07	5/8	87		
Automatic Cut off Single Cylinder		60			−80			.05	1/4	2.13		
Compound	9½ and 14½ by 14½		60	−31		−31	−113	.05	7/16		136	148
	10½ . 16 . 12		60	−31		−31	−113	.05	7/16		136	148
	12 . 17½ . 13¾		60	−31		−31	−105	.05	7/16		136	134
Corliss. Single Cylinder		60			−84			.03	1/4	2.21		
		60			−103			.03	1/6	2.56		
Compound	10 and 15 by 30		60	−61		−61	−136	.03	1/3		180	145
	12 . 18 . 36		60	−61		−61	−136	.03	1/3		180	145
	14 . 22 . 36		60	−61		−61	−143	.03	1/3		180	154
	14 . 22 . 42		60	−61		−61	−143	.03	1/3		180	154

COMPRESSED AIR.

Size of Engine Inches	Revolutions per Minute	Piston Speed per Minute Feet	Cubic Feet of Free Air per Minute	Brake Horse Power H.P.	Point of Cut off (Stroke = 1)	Temperature Fahrenheit in both Cylinders.		Gauge Pressure Lbs per Square Inch.		
						Initial	Final	Initial	Intermediate	Final
9½ and 14 by 24	112.5	450	444	50	.52	300	170	70	22.9	2
12 and 18 by 30	90	450	888	100	.52	300	170	70	22.9	2
13¼ and 20 by 30	90	450	1110	125	.52	300	170	70	22.9	2
13¼ and 20 by 36	75	450	1332	150	.52	300	170	70	22.9	2

— Cubic Feet of Free Air at 14.7 Lbs absolute and 60° Fahrenheit consumed per Minute in Corliss Compound Pneumatic Motors properly operated. —

— Remarks. — Brake Horse-Power is taken as .85 of Indicated Horse-Power. —

— Clearance in both Cylinders is assumed to be 3% of Theoretical Cylinder Capacity. —

— Cylinders jacketted for Hot Air. —

FIG. 20.

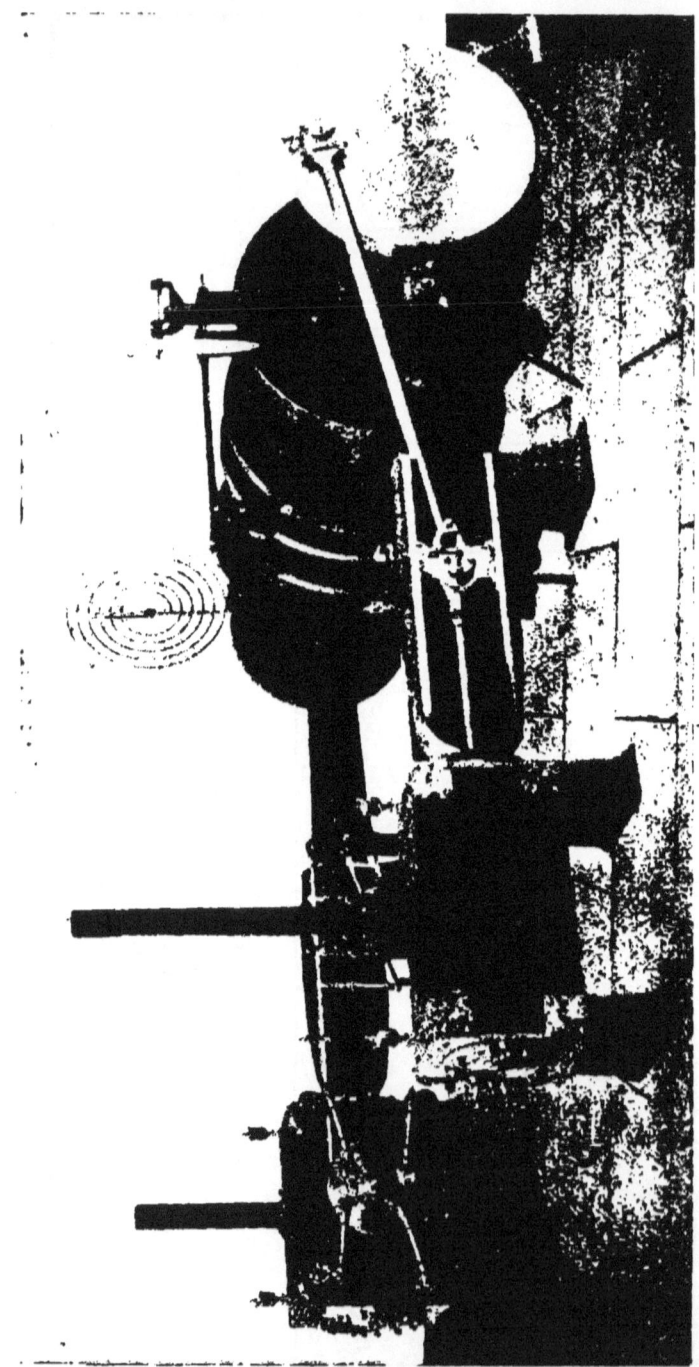

FIG. 20½.—Rix Pneumatic Direct Connected Hoisting Engines.
Manufactured by Fulton Engineering and Shipbuilding Works, San Francisco, Cal.

Mode of Compression		10	20	30	40	50	60	70	80	90	100	125	150	175	200
Adiabatic	A	2.8	16	12	9.8	8.5	7.54	6.8	6.36	5.9	5.6	4.94	4.5	4.13	3.9
	S	25.2	14.4	10.8	8.8	7.7	6.8	6.15	5.7	5.33	5	4.45	4	3.7	3.5
Single Stage Jacketed Cyl^s	A	28.2	16.3	12.2	10	8.7	7.7	7	6.6	6.2	5.8	5.1	4.7	4.3	4.1
	S	25.4	14.7	11	9	7.8	6.9	6.4	5.94	5.6	6.2	4.6	4.2	3.9	3.7
Two Stage Compound Jacketed	A	28.7	16.8	12.6	10.5	9.2	8.2	7.5	7	6.6	6.2	5.6	5.07	4.7	4.4
	S	25.8	15.12	11.34	9.45	8.3	7.4	6.75	6.3	5.85	5.6	5	4.56	4.2	4
Isothermal	A	30.3	18.2	14.1	11.9	10.5	9.6	8.9	8.4	8	7.6	6.9	6.45	6.1	5.8
	S	27.3	16.4	12.7	10.7	9.45	8.64	8	7.56	7.2	6.84	6.2	5.8	5.6	5.2

Receiver Gauge Pressure (Lbs per square inch).

Cubic Feet of Free Air at 60° Fahrenheit and 14.7 Lbs absolute per Square Inch compressed and delivered per Minute and per Indicated Horse Power.

Remark. — Letter A refers to the Air Cylinder. Letter S refers to the Direct-Acting Steam Cylinder.

FIG. 21.

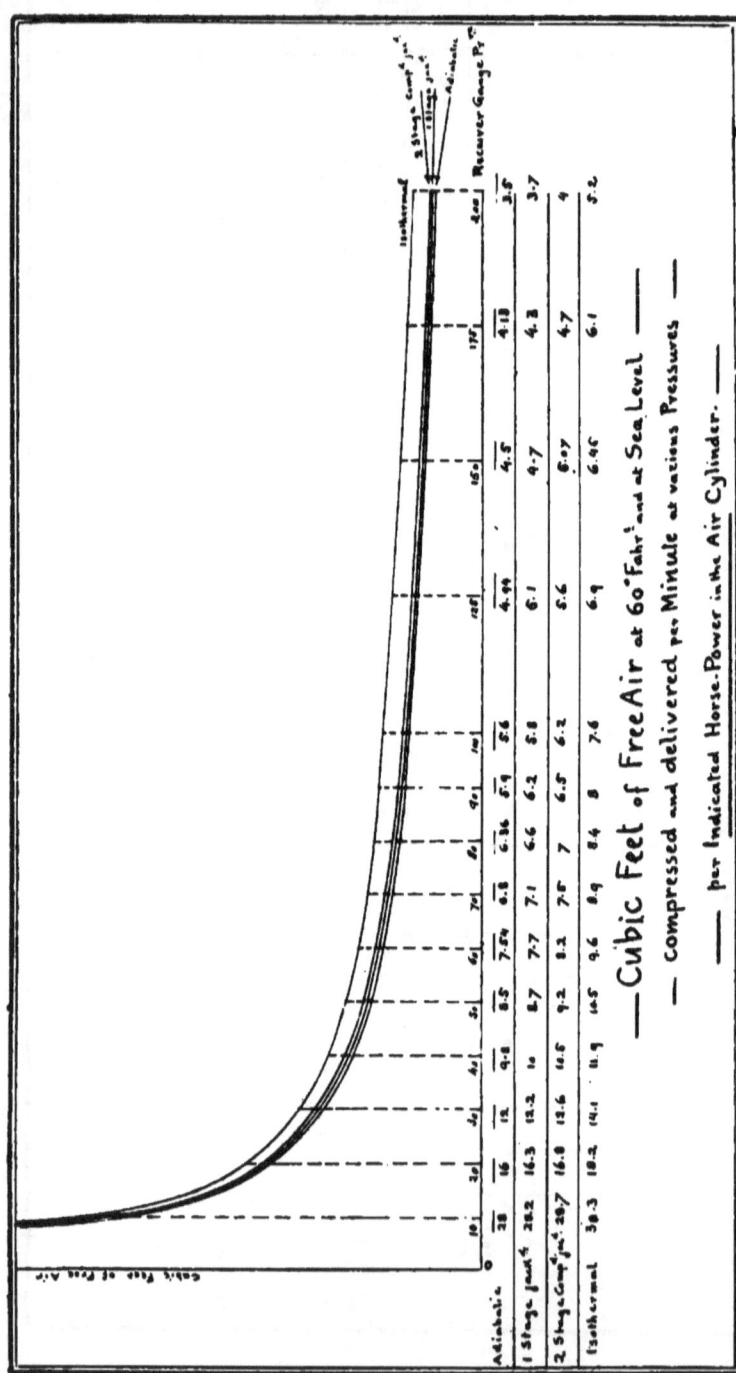

Fig. 22.

This table is constructed from the curve represented in Fig. 22.

In the table the amount of air is given for each indicated horse power in the air cylinder and also for each I. H. P. in the direct-acting steam cylinder which drives the compressor.

The table in Fig. 23 is practically the reverse of the preceding curve and the table gives the I. H. P. to compress and deliver 100 cubic feet per minute of air, at 60 degrees Fahr. and 14.7 lbs. per square inch absolute pressure. This table is constructed from the curve (Fig. 24) and gives the I. H. P. in the adiabatic compression, in single stage jacketed cylinder compression, in two-stage compound jacketed compression and also isothermal compression, and the horse powers under each of the different gauge pressures read both for the I. H. P. in the air cylinder and the I. H. P. in the direct-acting cylinder.

Fig. 25 is the curve of mean effective pressures per square inch in adiabatic compression, for the various receiver pressures enumerated. This will be found useful in computing piston loads.

Fig. 26 is a table of pressures per square inch, due to the weight of air at 60 degrees Fahr. in vertical pipes, and also the weight of one cubic foot of air in pounds avoirdupois. For example, if the gauge pressure at the surface of a mine is 70 lbs. per square inch, at the depth of one thousand feet the pressure will be 73 lbs. to the square inch. Where there are extreme variations in altitude in a transmission plant this weight of air has to be taken into consideration.

COMPRESSED AIR.

Mode of Compression		\-\-\- Receiver Gauge Pressure. (Lbs per square inch). \-\-\-												
		10	20	30	40	50	60	70	80	90	100	125	150	175	200
Adiabatic	A	3.57	6.22	8.33	10.21	11.73	13.25	14.67	15.72	16.88	17.83	20.21	22.31	24.18	25.74
	S	3.93	6.84	9.16	11.23	12.9	14.57	16.14	17.29	18.57	19.61	22.23	24.54	26.6	28.31
Single Stage Jacketed Cyln	A	3.55	6.14	8.21	9.96	11.46	12.84	14.07	15.18	16.22	17.2	19.33	21.22	22.9	24.42
	S	3.9	6.75	9.03	10.95	12.6	14.12	15.48	16.7	17.84	18.9	21.26	23.34	25.2	26.86
Two Stage Compound Jacketed	A	3.48	5.95	7.91	9.48	10.87	12.2	13.25	14.27	15.29	16.06	18.05	19.72	21.23	22.5
	S	3.83	6.54	8.7	10.43	11.96	13.4	14.57	15.7	16.82	17.67	19.83	21.69	23.35	24.7
Iso-thermal	A	3.3	5.5	7.1	8.4	9.5	10.4	11.2	11.9	12.6	13.2	14.4	15.5	16.4	17.2
	S	3.6	6.1	7.8	9.2	10.4	11.4	12.3	13.1	13.8	14.5	15.8	17	18	18.9

\-\-\- Indicated Horse-Power to Compress and deliver 100 Cubic Feet per Minute of Air at 60° Fahrenheit and 14.7 Lbs per square inch absolute pressure. \-\-\-

Remark. \-\-\- Letter 'A' refers to the Air Cylinder. Letter 'S' refers to the Direct-Acting Steam Cylinder.

FIG. 23.

COMPRESSED AIR.

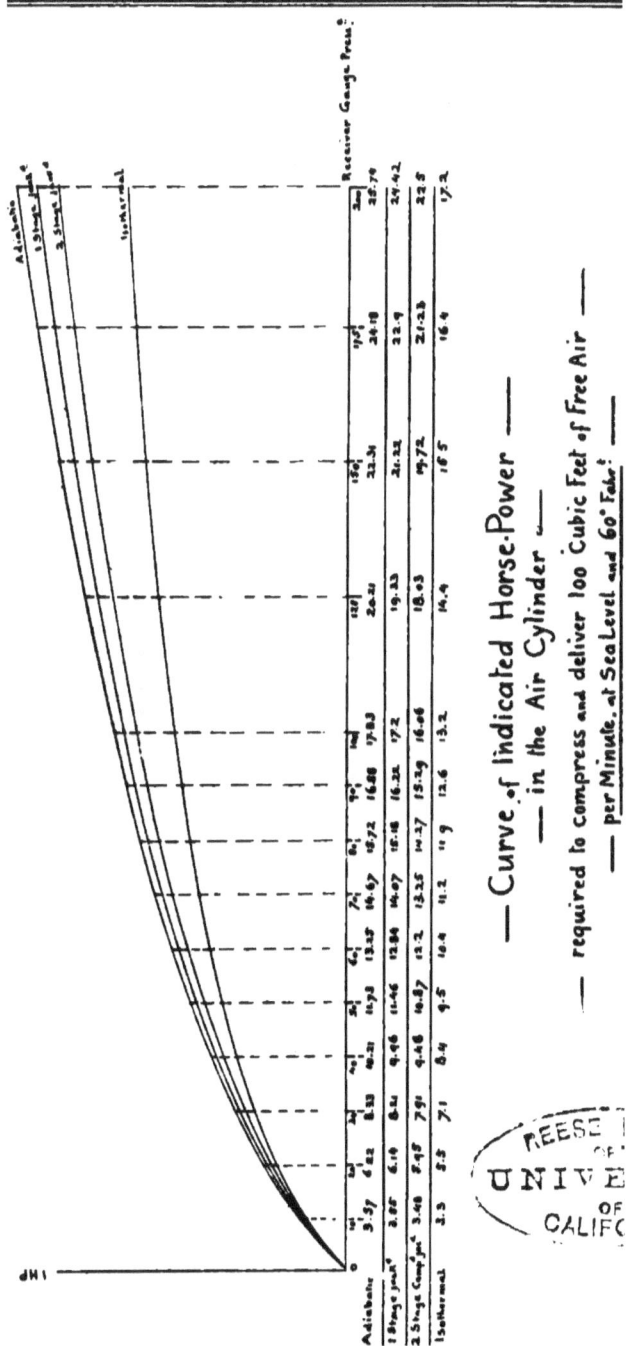

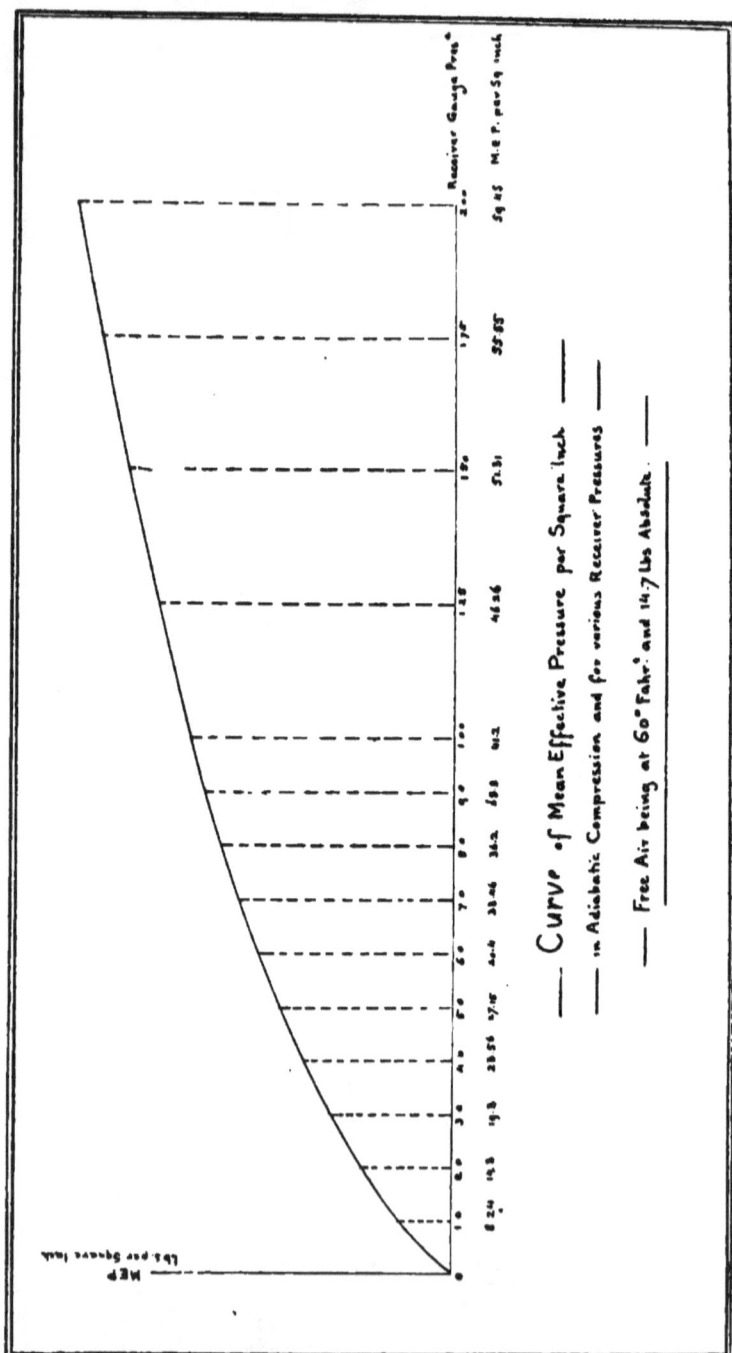

Fig. 25.

Gauge Pressure Lbs per Square Inch	50	60	70	80	90	100
Weight of One Cubic Foot of Air Lbs	.336	.388	.44	.492	.544	.596
Weight per Square Inch for 100 ft Vertical (Lbs)	.23	.27	.3	.34	.38	.41

— Pressure per Square Inch due to the Weight of Air at 60° Fahrenheit —
— in Vertical Pipes, the Atmospheric Air being at 14.7 Lbs absolute, per Square Inch. —

FIG. 26.

AMOUNT OF FREE AIR REQUIRED TO RUN DIRECT-ACTING STEAM PUMPS.

In preparing these tables the object has been to furnish information to the oft-repeated query, "How many cubic feet of free air, compressed to, say 60 lbs., is required to run a direct-acting pump that will raise 50 gals. per minute 200 feet high, or say 8 miners' inches 150 feet high, or at any other pressure of air?" We have made three assumptions in these calculations, which are likely to cover all possible losses of efficiency in ordinary work.

First—The work absorbed by the pump has been estimated by adding 20 per cent to the actual work in water raised, to make up for frictional and other resistances.

Second—The actual capacity of the air cylinder, that is, the volume swept by the piston, has been increased by 15 per cent to take into account the clearance, leakage, etc.

Third—The working pressure of air, when entering the air cylinder, has been taken at 10 lbs. per square inch lower than the receiver pressure, to compensate for frictional and other resistances.

We have not assumed that the air was reheated before entering the cylinder, nor was any account taken of the difference of level between the receiver and the pump, which in many cases would add several pounds per square inch to the working pressure, as noted in the Table (Fig. 26). The results given in these tables may therefore be referred direct to the intake capacity of the compressor and the estimate of the air consumption required is therefore very much simplified.

If the necessary power to produce the quantities of compressed air indicated in these tables be compared to the corresponding work in water raised, the efficiency, which is measured by the ratio of the latter to the former, will be as low as 25 per cent. A direct-acting pump does not use air expansively, and this is well known to be a simple but a wasteful manner of transmitting power.

Assuming the values in these tables to be one, the following table will show the percentages required for the different kinds of power-actuated pumps, both for cold air and air delivered at 300 degrees Fahr. at the pump motor.

Kind of Motor.	AIR.	
	Cold. (60° F.)	Reheated to 300° F.
Direct Acting Single...............	1	.69
Direct Acting Compound...........	.70 to .60	.48 to .41
Fly Wheel { Slide Valve Single	.60	.41
Fly Wheel { Slide Valve Compound	.50	.329
Fly Wheel { Corliss Compound	.33	.226

Miners Inches	Head in Feet																							
	50			100			150			200			250			300			350			400		
	60 lbs	70 lbs	80 lbs	60 lbs	70 lbs	80 lbs	60 lbs	70 lbs	80 lbs	60 lbs	70 lbs	80 lbs	60 lbs	70 lbs	80 lbs	60 lbs	70 lbs	80 lbs	60 lbs	70 lbs	80 lbs	60 lbs	70 lbs	80 lbs
1	4.4	4.24	4.1	8.8	8.46	8.2	13.2	12.7	12.3	17.6	16.94	16.4	22	21.2	20.5	26.4	25.4	24.6	30.8	29.7	28.7	35.2	33.9	32.8
2	8.8	8.5	8.2	17.6	17	16.4	26.4	25.4	24.6	35.2	33.9	32.8	42.2	42.4	41	52.8	50.8	49.2	61.6	59.4	57.4	70.4	67.8	65.6
3	13.2	12.7	12.3	26.4	25.5	24.6	39.6	38.1	36.9	52.8	50.9	49.2	66	63.6	61.5	79.2	76.2	73.8	92.4	89.1	86.1	105.6	101.7	98.4
4	17.6	17	16.4	35.2	34	32.8	52.8	50.8	49.2	70.4	67.8	65.6	88	84.8	82	105.6	101.6	92.4	123.2	118.8	114.8	144.8	135.6	131.2
5	22	21.2	20.6	44	42.9	41	66	63.6	61.5	88	84.8	82	110	106	102.5	132	127	123	154	148.5	143.5	176	169.5	164.0
6	26.4	25.5	24.6	52.8	51	49.2	79.2	76.2	73.8	105.6	101.8	98.4	132	158.4	123	158.4	162.4	147.6	184.8	177.2	172.2	211.2	203.4	196.8
7	30.8	29.7	28.7	61.6	59.3	57.4	92.4	89	86.1	123.2	118.7	114.8	154	148.4	143.5	184.8	178	172.2	215.6	207.9	200.9	246.4	237.3	229.6
8	35.2	34	32.8	70.4	68	65.6	105.6	101.6	98.4	140.8	135.6	131.2	176	169.6	164	211.2	203.2	196.8	246.4	237.6	229.6	281.6	271.2	262.4
9	39.6	38.1	36.9	79.2	76.5	73.8	118.8	114.3	110.7	158.4	152.4	147.6	198	190.8	184.5	237.6	228.6	221.4	277.2	267.3	258.3	316.8	305.1	295.2
10	44	42.4	41	88	84.8	82	132	127.2	123	176	169.6	164	220	212	205	264	254	246	308	297	287	352	339	328

—— Cubic Feet of Free Air (14.7 lbs. Absolute) per Minute ——
—— Compressed to 60 – 70 – 80 Lbs Receiver Gauge Pressure ——
—— To operate Direct-Acting Pumps working against various Heads ——
—— Remark — Pumps assumed to work at 100 feet Piston velocity per Minute ——

FIG. 27.

COMPRESSED AIR.

Gallons per Minute	Head in Feet																							
	50				100				150				200				250				300			
	60 lb	70 lb	80 lb		60 lb	70 lb	80 lb		60 lb	70 lb	80 lb		60 lb	70 lb	80 lb		60 lb	70 lb	80 lb		60 lb	70 lb	80 lb	
10	3.7	3.5	3.4		7.4	7	6.8		11.1	10.5	10.2		14.8	14	13.6		18.5	17.5	17		22.2	21	20.4	
20	7.4	7	6.8		14.8	14	13.6		22.2	21	20.5		29.6	28	27.2		37	35	34		44.4	42	40.8	
30	11.1	10.5	10.2		22.2	21	20.4		33.3	31.5	30.6		44.4	42	40.8		55.5	52.5	51		66.6	63	61.2	
40	14.8	14	13.6		29.6	28	27.2		44.4	42	40.8		59.2	56	54.4		74	70	68		88.8	84	81.6	
50	18.5	17.5	17		37	35	34		55.5	52.5	51		74	70	68		92.5	87.5	85		111	105	102	
60	22.2	21	20.4		44.4	42	40.8		66.6	63	61.2		88.8	84	81.6		111	105	102		133.2	126	122.4	
70	26.9	24.5	23.8		51.8	49	47.6		77.7	73.5	71.4		103.6	98	95.2		129.5	122.5	119		155.4	147	142.8	
80	29.6	28	27.2		59.2	56	54.4		88.8	84	81.6		118.4	112	108.8		148	140	136		177.6	168	163.2	
90	33.3	31.5	30.6		66.6	63	61.2		99.9	94.5	91.8		133.2	126	122.4		166.5	157.5	153		199.8	189	183.6	
100	37	35	34		74	70	68		111	105	102		148	140	136		185	175	170		222	210	204	

Gallons per Minute	350				400			
	60 lb	70 lb	80 lb		60 lb	70 lb	80 lb	
10	25.9	24.5	23.8		29.6	28	27.2	
20	51.8	49	47.6		59.2	56	54.4	
30	77.7	73.5	71.4		88.8	84	81.6	
40	103.6	98	95.2		118.4	112	108.8	
50	129.5	122.5	119		148	140	136	
60	155.4	147	142.8		177.6	168	163.2	
70	181.3	171.5	166.6		207.2	196	190.4	
80	207.2	196	190.4		236.8	224	217.6	
90	233.1	220.5	214.2		266.4	252	244.8	
100	259	245	238		296	280	272	

— Cubic Feet of Free Air (14.7 lb Absolute) per Minute —
— Compressed to 60 – 70 – 80 Lbs Receiver Gauge Pressure —
— to operate Direct-Acting Pumps working against Various Heads —
— Remark :— Pumps assumed to work at 100 Feet Piston Velocity per Minute. —

FIG. 28.

REFRIGERATION BY COMPRESSED AIR.

This Treatise would soon grow beyond reasonable limits if it had to enumerate all the applications of compressed air in modern industry; in fact, a publication claiming to give an exact "up to date" account of these applications would never come to an end, as some new and unexpected uses are constantly arising.

But a review, however cursory, of the properties of compressed air considered as motive power, must of necessity touch upon one of its most interesting uses; viz., the production of cold. A rapid treatment of this question is here the more justified as it does not correspond to a class of apparatus intended solely for refrigerating purposes.

Every air motor is in itself, and at no additional cost, a cold-producing machine, and this property, which belongs exclusively to compressed air, will often be found a valuable addition to its other merits, especially by the underground worker.

A quotation from Prof. A. B. W. Kennedy on the Paris air installations may fitly be reproduced here:

"By using air direct from the main in the motor, or by heating it only very slightly, the exhaust air can be, of course, so greatly reduced in temperature as to be available for freezing purposes.

"In one Paris restaurant, for instance, which I visited, I found that the exhaust was carried through a brick flue into the beer cellar. In this flue the carafes were set to freeze, large molds of block ice were also being made for table use, while the air was still cold enough in passing away through the beer cellar to render the use of ice for cooling quite unnecessary even in the hottest weather.

"The nominal function of the engine in this case was the charging of batteries used in the electric lighting of the restaurant.

"The conjoint use of power and cold is common in Paris, the power being in this case generally applied to electric lighting. While in any large city, such as Paris, it is no doubt a great point that by a compressed air system the handiest possible cooling appliances can be brought everywhere within reach, in tropical climates this is something rather of necessity than of luxury. In such cases we might have the apparent paradox of a motor worked essentially for its exhaust; the work done would be a bye-product, the cold air would be the principal thing.

"In such a case, if there were no useful work to be done, the motor could even be made (as has been suggested to me) to pump air back into the main, and thus to virtually halve its air consumption."

From these remarks the conclusion is obvious that ice-making, water-cooling, and cold-storage contrivances are of easy application whenever air motors are used; and it will be readily understood that the exhaust temperature of air may be regulated by a variation in the degree of heating.

An inexpensive and tolerably efficient arrangement consists of exhausting the air from the motor at one end of a duct made of insulating material, such as two or more parallel courses of one-inch boards, paper-coated on the outside, and secured one or two inches apart by wooden strips; or, else, in a more permanent installation the duct may be a brick flue, such as described in the above report.

In both cases its upper portion can be laid open, and arrangements are made at the interior of it for suspending ice molds, water pails, etc., which are removed at intervals depending upon the exhaust temperature of the motor and its activity.

Provision should be made to rid the exhaust air from all the grease or oil which it might carry out of the motor before it is admitted into the duct.

One noteworthy feature about air thus used for cooling purposes is its wholesome nature; with its defects, adiabatic compression is endowed with this beneficial property that the combined heat and pressure thus generated prove too much for the endurance of microbes; air thus treated becomes thoroughly sterilized, and can be safely put in contact with alimentary substances at no risk of contamination. In fact, fruits are wonderfully preserved during transportation by a new system wherein the exhaust from the air brake cylinders is the vital principle.

More elaborate ice-making or cold-storage appliances might, of course, be devised, and special machinery has been constructed to that effect.

It is not here intended to treat upon the general subject of ice-making machines. This cannot be done in an elementary way with any degree of completeness. Referring solely to the air machine, it may be stated that it is not the most economical for cold production, but in many instances it remains in use because of its convenience and safety.

Air is found everywhere, and in case of leakage is not apt, like ammonia or sulphur dioxide, to spoil the provisions subjected to cooling.

The developments previously expounded in this treatise will facilitate the comprehension of the cold air machine, which is principally used on shipboard, for the preservation of provisions, and for ice-making.

Two principal types of machines are commonly used.

THE BELL-COLEMAN MACHINE.

A revolving shaft d, is operated either by a steam engine S, and a crank c, as shown on the line drawing, or by a belt transmission.

This shaft carries two flywheels W, W, and two opposite cranks e, f, actuating by connecting rods and crossheads two pistons v, h.

COMPRESSED AIR. 65

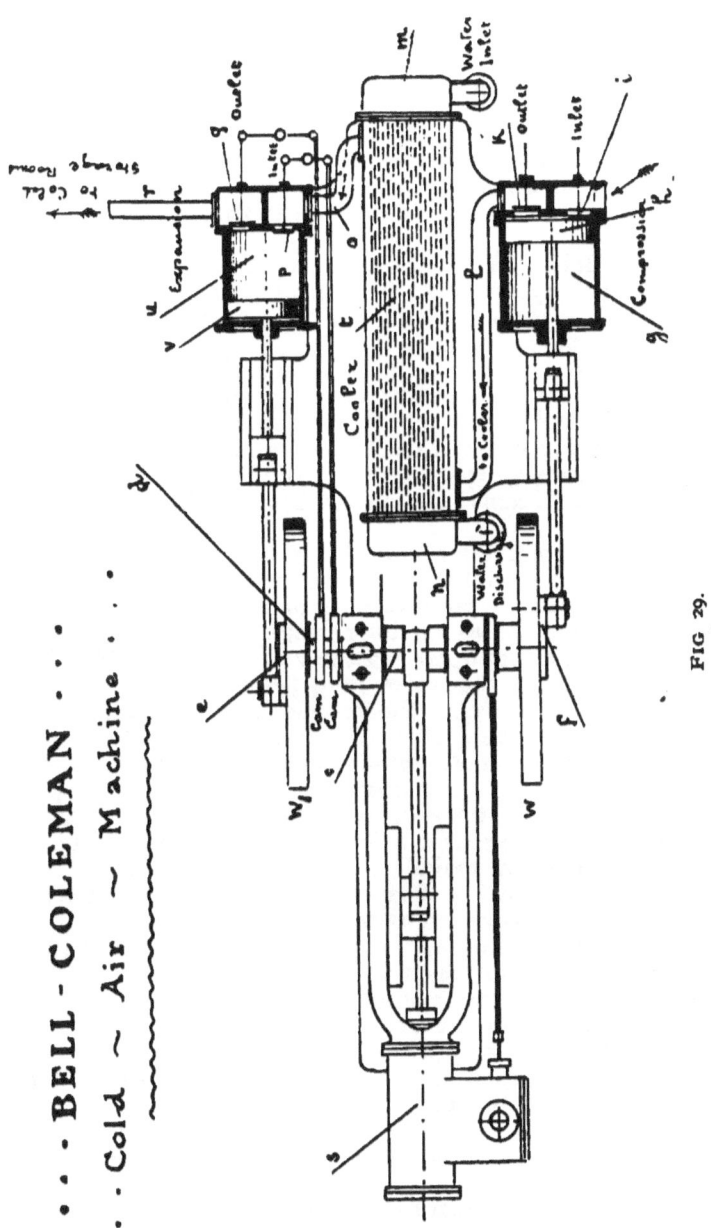

FIG. 29.

The piston h travels in a cylinder g, which, for the sake of simplicity, has been shown single-acting. The piston h is solid, and the back head of the cylinder carries two automatic valves i, K. The valve i, opening inward, is an inlet valve admitting the free air into the cylinder, during one stroke of the piston h; during the reverse stroke the valve i is closed, the air confined in the cylinder is compressed, and escapes through the outlet valve K, and the pipe l, into a tubular cooler, through which a series of tubes t, establishes a continuous circulation of cold water. This water enters the cooler through the cover m, and is discharged through the opposite end n.

The air delivered by the compressing cylinder g, passes around the tubes t, is cooled to, or nearly to, the outside temperature, and passes through the pipe o, connected to the backhead of another cylinder u.

This cylinder, which is also shown single-acting, has a solid piston v, operated by the crank e; the backhead carries two separate and closed chambers, containing one an inlet valve p, and the other a discharge valve q; but instead of acting automatically, these valves have their motion controlled by two adjustable cams revolving within the shaft d, as shown on the cut.

While the compression cylinder g delivers at each stroke some compressed air into the cooler, the inlet valve p admits into the cylinder q a certain volume of this air, which, as said before, has been cooled on its passage around the tubes, but the setting of the cam operating the valve p on the shaft is so arranged as to close this valve long before the piston v is at the outer end of its stroke; the volume of air introduced into the cylinder u is then left to expand adiabatically, and its temperature falls to a point which depends upon the amount of expansion, i. e. upon the quantity of air admitted by the valve p; besides, this work of expansion helps the motion of the machine to some extent. For this reason, the cylinder u is called the expansion cylinder.

When the piston v has reached the end of its stroke, the discharge valve q is opened by its cam, and so remains during the whole reverse stroke, the piston v driving the cold air through the pipe r, to the cold storage rooms.

It will be readily understood that when the valve p closes early on the stroke of the expansion piston v, the pressure in the cooler increases, and the exhaust temperature in the cylinder u decreases; when, on the contrary, the closing of the valve p is retarded, the pressure in the cooler drops, and the exhaust temperature rises. So that, by a proper adjustment of the cams, the degree of cooling air can be varied within a large range.

This machine is comparatively cumbersome, if the amount of cooling is important.

THE ALLEN DENSE-AIR MACHINE.

To obviate this defect, another class of machines has been devised, known as the ALLEN DENSE-AIR MACHINE. Its general arrangement being practically the same, no special drawing of it is given.

The air penetrating into the compression cylinder has been primitively raised to a certain pressure, say 40 lbs.; the compression carries this pressure to, say 160 lbs.; then the air passes through the cooler into the expansion cylinder, wherein it again expands from 160 lbs. to 40 lbs. or more, according to the temperature at which it is desired to discharge it through a pipe like *r;* but instead of being allowed to diffuse freely into the cold storage ducts or chambers, this air is circulated through coils of closed pipe, which are finally connected to the inlet valve chamber of the compression cylinder, the same air being thus used over and over again.

This machine operates upon a greater weight of air under a given volume, and consequently is more effective under this volume, than the Bell-Coleman machine. Or else, the Allen machine can produce the same cooling effect with smaller dimensions than the Bell-Coleman, which is an important feature on shipboard.

It is interesting to form an idea of the practical results which can be attained with this sort of machine, to which object the following data refers:

One pound of ice at 32 degrees to be transformed into water at 32 degrees, absorbs 142 B. T. U. without variation of temperature.

This amount of heat, which disappears, without influencing the thermometer's indications, is termed the latent heat of fusion of ice. In the same way, if we want to transform 1 lb. of water at 32 degrees Fahr., into ice at 32 degrees Fahr., we must subtract 142 B. T. U. from that pound of water, without changing its temperature, and these 142 B. T. U. are the latent heat of solidification of water. These two terms are entirely equivalent.

One ton of 2000 lbs. of ice in the process of melting into water at 32 degrees Fahr., will therefore subtract from the surrounding bodies, air, water, or whatever they may be, 2000×142 or 284,000 B. T. U., and the resulting effect produced on those bodies is measured by *1 ton of ice melting capacity*.

The refrigerating action of a machine or process of any kind, producing that same effect, is estimated in the same terms, and such a machine is said to have a cooling capacity of *1 ton ice* melting.

The annexed Table gives the numbers of negative B. T. U. and of lbs. of ice melting capacity, developed in the adiabatic expansion of 1 cubic foot of air from 60, 70, 80, 90, and 100 lbs. gauge respectively to 14.7 lbs. absolute; these are the calculated or theoretical capacities.

These figures show that there is no advantage in using a high air pressure, because the refrigerating capacity does not

Initial Gauge Press. Lbs per Square Inch	Initial Temperature Fahrenheit	Final Temperature Fahrenheit	Range of Cooling Degrees Fahrenheit	Negative B.T.U. Calculated	Negative B.T.U. Practical	Lbs of Ice Melting Capacity Calculated	Lbs of Ice Melting Capacity Practical
60	60	-135.8	195.8	3.524	1.515	.0248	.0103
70	60	-147.5	207.5	3.735	1.600	.0263	.0113
80	60	-157.4	217.4	3.913	1.682	.0275	.0118
90	60	-166.2	226.2	4.072	1.751	.0286	.0123
100	60	-173.8	233.8	4.208	1.809	.0296	.0127

Refrigerating Capacity of One cubic Foot of Compressed Air expanded to 14.7 Lbs Absolute
Outside Temperature: 60°Fahr.

Remark.—The Practical Capacities are determined from experiments on Bell-Coleman Engines.

FIG. 30.

vary proportionately with the rise of pressure; for instance, if this latter passes from 60 to 70 lbs., the pressure increases by 16 per cent and the cooling capacity by 9 per cent; while if the pressure becomes 100 lbs. the increase of pressure is 66 per cent and the increase of cooling capacity is 23 per cent. In other words, the percentage of pressure having increased 4.1 times, the percentage of cooling capacity rises only 2.5 times.

No very complete record of experiments has been published showing the practical efficiency of cold air machines, one reason being that, for this special object, their use is limited, as compared with the ammonia machines. But whenever cold air machines are adopted, it is because their efficiency is superseded by other practical reasons. Some accurate tests place it at 43 per cent of the theoretical cooling capacity of the air for the Bell-Coleman, and 37 per cent for the Allen Dense-Air Engine.

The former coefficient has been used to establish the column of the Table headed "Practical Capacity."

But while in a large city, where ammonia can be readily obtained a wholesale ice-making and cold-storage business would not be undertaken with a compressed air plant, it is none the less certain that in mining camps, or in remote localities, where cheap motive power is frequently obtainable, the application of air to the production of cold remains one of the most interesting and profitable adjuncts of this valuable power agency.

FIG. 31.—Pneumatic Locomotive.

POWER TRANSMISSION BY COMPRESSED AIR.

The use of compressed air for transmitting power for a long distance is daily gaining in importance, and this application of air, which a few years ago did not receive much consideration, outside of actuating rock drills and coal cutters, stands at present as one of the most economical and satisfactory systems of power transmission.

No fair-minded and impartial person will contend that in every possible case in practise one particular system of power transmission can be always preferable to all others.

The economical solution of industrial problems involves so many factors of entirely independent and often contradictory nature, that the strictly engineering side of the question may be overcome in importance by other conditions which would hardly have been thought of at a first glance.

One fact however, may be stated as general, and that is that compressed air is better adapted to underground work than any other agency of power transmission. Not only can it be transported anywhere, through crooked and narrow passages, either wet or dry, and regardless of insulation or losses other than leakage at the pipe joints, but its use and handling is totally devoid of danger, and after its work is done it becomes the most essential element to human life. This can be said of air alone, although it has nothing to do with its value as a power transmitter.

The principles of the production of compressed air are expounded in another part of this Treatise; it is therefore unnecessary to explain here, how, when air has been raised to a certain pressure, with production of heat, and when on its passage through a long pipe, this air has cooled down to the temperature of the atmosphere, the loss of efficiency incurred in this drop of temperature can be balanced, and even exceeded, by reheating the air before it is admitted to the motors.

This reheating, which can be done at a small expenditure of fuel, is an important element in the total efficiency of the system.

Being now in possession of all the essential elements of information required in estimating the size and the cost of a compressed air transmission, their application to some practical examples will form a fitting conclusion to the preceding developments:

EXAMPLE 1.

A stamp mill is located at 3000 feet from a water wheel developing 70 B. H. P.

Required a compressed air transmission to deliver 40 H. P. on the line shaft.

The motor operating the mill is 500 feet higher up than the air receiver, wherein the air pressure is to be 80 lbs.

Altitude of compressor: 3500 feet above sea level.
Temperature at compressor and Mill, 50 degrees Fahr.

The loss from belt-slipping between the cam shaft and the motor shaft, and from other causes, can be taken as 10 per cent, and the I. H. P. of the motor will be:

$$\frac{40}{9} = 44.4$$

and assuming another loss of 8 per cent for clearance, wire drawing, etc., the available power at the lower end of the main must be: 48.3 H. P.

As there is, at first glance, an important margin between the powers at the wheel and at the mill, it naturally occurs to consider whether the reheating of the air at the motor cannot be dispensed with.

We will use a slide-valve engine, cutting off at ½ stroke, which would likely be the earliest admissible cut-off as regards exhaust temperature.

The amount of air necessary to develop 48.3 H. P. is: 1.13 lbs. per second, or 67.8 lbs. per minute. If we were at the sea level, 1 lb. of air at 60 degrees Fahr. would represent 13.1 cubic feet.

Sixty-seven and eight-tenths pounds represent, therefore: $67.8 \times 13.1 = 888.2$ cubic feet.

If we use a single stage compressor, the I. H. P. required for 80 lbs. gauge receiver pressure will be:

$$15.18 \times 8.882 = 134.83.$$

But the Table of columns and powers at various altitudes shows that the power required to compress and deliver the same volume of air at the same pressure, but at 3500 feet altitude, is (Cols. 5 and 6):

$$\frac{14.93}{.89 \times 15.72} = 1.067 \text{ times greater than at the sea level.}$$

And as the temperature is 50 degrees Fahr., this figure should be reduced in the ratio of the absolute temperature (at 60 degrees and 50 degrees Fahr.) and becomes 1.046.

The power actually required will therefore be:

$$134.83 \times 1.046 = 141 \text{ I. H. P. in the compressor.}$$

And if we allow it mechanical efficiency, the brake power on the wheel is:

$$\frac{141}{.9} = 155 \text{ B. H. P.}$$

Whilst we have only 70 B. H. P. at our disposal.

The air cannot therefore be used cold in the motor; in other words, we have not yet a sufficient margin of power between the wheel and the mill to permit the use of cold air; reheating must necessarily be resorted to.

We have 70 B. H. P. on the compressor shaft, and $70 \times 0.9 = 63$ I. H. P. in the air cylinder.

From the above calculations, we know that the compression and delivery of 100 cubic feet of free air per minute at 80 lbs. receiver pressure, and at the given altitude and temperature, require: $15.18 \times 1.046 = 15.88$ I. H. P.

COMPRESSED AIR.

The available power of 63 I. H. P. will permit of compressing $100 \times \frac{63}{15.88} = 397$ cubic feet of free air per minute, whose weight at 3500 feet altitude and 50 degrees Fahr. is:

$$397 \times .0807 \times \frac{493}{511} = 30.97.$$

Giving per second a weight of air of:

$$\frac{30.97}{60} = .516 \text{ lbs.}$$

We have next to determine the air pressure at the lower end or outlet of the main, for a length of 3000 feet.

The tables of frictional resistance show that 80 lbs. gauge (94.7 lbs. absolute) being the pressure at entrance to the main, the pressure at the lower end is:
With a 4-inch main: 77.7 lbs. gauge.
With a 3-inch main: 73. lbs. gauge.

Besides, as the outlet of the main is 500 feet above the receiver, we lose from this fact 1.7 lbs., which leaves as available pressures at the outlet:
With a 4-inch main: 76 lbs. gauge.
With a 3-inch main: 71.3 lbs. gauge.

We will use the 4-inch main, and as the necessary reheating obviates the low temperature of exhaust caused by a long expansion, we will use a motor expanding from 76 lbs. to 2 lbs. gauge, and find that to develop 48.3 H. P. with 516 lbs. of air per second, this air must be reheated to 247 degrees Fahr.

What amount of fuel this reheating will require can be easily computed.

We have to reheat 30.97 lbs. of air per minute, from 50 degrees to 247 degrees Fahr., or 197 degrees Fahr.

The specific heat of air being .238, this will require:
$30.97 \times .238 \times 197 = 1451.9$ B. T. U. per minute, or:
$1451.9 \times 1440 = 2,090736$ B. T. U. per 24 hours.

And allowing that 1 lb. of coal will yield 10,000 B. T. U. 209 1 lbs. of coal per 24 hours.

Or if 1 lb. of pine wood will yield 5400 B. T. U., the weight consumed per 24 hours is: $309.1 \times \frac{10\,000}{5,400} = 386.84$.

Or about ⅕ cord.

SIZE OF COMPRESSOR.

We found as the "useful" amount of air per minute 397 cubic feet, and allowing .85 volumetric efficency for the compressor, its intake capacity must be 467 cubic feet.

With 300 feet per minute piston velocity, and referring to Table (Fig. 37), the compressor will be a single 18½-inch machine, or a duplex 12½-inch machine.

EFFICIENCY OF THE TRANSMISSION.

The apparent efficiency of the transmission is:

$$\frac{40}{70} = .57$$

But its exact value should take into account the coal consumed in reheating the air.

This latter amounts to 8.70 lbs. per hour, and if we assume that in a compound steam engine the coal consumption is 2 lbs. per I. H. P., this quantity represents: $\frac{8.70}{2}=4.35$ I. H. P. on the piston of a direct-acting steam engine operating the compressor, or, in the present case, on the compressor shaft.

The true efficiency is therefore:
$$\frac{40}{74.35}=.54$$
With reverse conditions, i. e., mill 500 feet below compressor, the total efficiency would be 55.

This is an example of a comparatively low efficiency in transmission. The power is so small that comparative losses become large. This transmission, however, can be improved by using a 2-stage compound compressor and motor, the calculations for which would be as follows:

379 cu. ft. of free air at sea level, and 50° Fahr. 4″ pipe.

Absolute pressure { At entrance, 114.7 lbs. (100 g.)
{ At exit, 114 lbs. (99.3 g.)

COMPOUND MOTOR.

First reheating, 50° to 350° Fahr.
    Power in H. P. cylinder.................. 32.84
Second, 153° to 350° Fahr.
    Power in L. P. cylinder.................. 27.26

                                 Total................. 60.10

First loss, 8 per cent, as in preceding example.
                60.1 × .92 = 55.29
Second loss, 10 per cent
                55.29 × .9 = 49.76

                                      on line shaft.

Coal used for reheating: 12.58 lbs. per hour, corresponding to: 6.29 I. H. P.

Total efficiency: $\frac{49.76}{76.29}=\underline{652}$

                                  65.2 per cent.

EXAMPLE 2.

A system of Power Transmission will now be considered in the case of a large mine, requiring:

100 B. H. P. for hoisting   ⎫
100 B. H. P. for pumping ⎬ At the surface.
100 B. H. P. for a stamp mill ⎪
25 B. H. P. for lighting   ⎭

And

25 B. H. P. for hoisting   ⎫
25 B. H. P. for pumping   ⎬
1500 cu. ft. of free air per ⎬ At 1500 ft. level.
minute at 60 degrees Fahr. ⎪
for rock drills   ⎭

COMPRESSED AIR.

Length of Transmission to surface plant: 4 miles.
Compressors of the 2-stage Compound type.
Receiver pressure 75 lbs. gauge.
Outside temperature 60 degrees Fahr.
Permissible loss of pressure:
        In surface main: 1 lb.
        In underground: ½ lb.
Required:
Size of surface and underground mains.
Size of Compressor.
B. H. P. on compressor shaft.

The power received at the mine will be divided under two heads, viz: *Surface* and *Underground*.

SURFACE PLANT.

We will assume for the motors a mechanical efficiency of .9, giving $\frac{325}{.9} = 361$ H. P.; and then, another loss of 5 per cent between the cylinder and the lower end of the main, for wire-drawing, elbows, etc.

The available power at the end of main is:
$$\frac{361}{.95} = 380,$$
the total efficiency being $.9 \times .95 = .855$.

The absolute pressure at upper end of main is: 89.7
The absolute pressure at lower end of main is: 88.7
and the weight of air at this pressure, reheated to 400 degrees Fahr. and completely expanded is: 3.26 lbs. per second.

UNDERGROUND PLANT.

Pressure at top of main: 88.7
Pressure at bottom of main 88.2
Additional pressure at bottom of main, 4.8, due to weight of air.
Absolute pressure at 1500 level 93.0

Fifty B. H. P. with .855 efficiency give: 58.5 H. P., which require a weight of air per second of: .59 lbs.

The rock drills work practically at full pressure, and the expansion of 19 lbs. (to 60 lbs. gauge) cannot be utilized.

The temperature of the compressed air at the bottom of shaft column is 250 degrees, and assuming a loss of 100 degrees before reaching the drills, i. e., a temperature of 150 degrees at the drills, the 1500 cubic feet of air will have to be reduced in the ratio of: $\frac{621}{611}$, and become: 1275 cu. ft., whose weight is (per minute) 97.41 lbs.

The total weight of air to be supplied per second is, therefore:

    Surface: 3.26
    Underground: .59
    Rock Drills: 1.62
        Total   5.47 lbs.

corresponding to 4299.43 cubic feet per minute of free air at 60 degrees Fahr., whose compression, in a 2-stage Compound Compressor, will require:

$$578.8 \text{ I. H. P.}$$

With .9 mechanical efficiency, the B. H. P. is:

$$643 \text{ B. H. P.}$$

The reheating will require 159.34 lbs. of coal per hour, corresponding to: 67 B. H. P., making a total B. H. P. of, 710 B. H. P.

The power obtainable at the lower end of main is:

$$637.3 \text{ H. P.}$$

and on the shaft of the motors, with .855 efficiency:

$$545 \text{ B. H. P.}$$

The total actual efficiency is, therefore: $\frac{545}{710} = 76.7$.

SIZE OF MAINS.

The Tables of frictional resistances, of which the use has been explained, give, as proper size of the pipes:

For the surface main: 12½ inches.
For the shaft column: 6⅝ inches.

SIZE OF COMPRESSOR.

The useful capacity has been found as:

4299.42 cubic feet of free air per minute.

Taking for the compressor 85 per cent volumetric efficiency, the actual capacity is: 5058 cubic feet, and with 400 feet of piston velocity, we will find by referring to Fig. 37 the proper size of the compressor.

It is desirable, for reasons of practical convenience, to divide the compressing plant in two equal units, each formed of a duplex compound machine. There will be, consequently, 4 intake cylinders, each having a capacity of 1264 cubic feet per minute, which correspond to a diameter of 24½ inches.

For 75 lbs. gauge receiver pressure, the area of the H. P. cylinder should be: 189.36 square inches, corresponding to 15½ inch bore, and at the rate of 80 revolutions per minute, the stroke will be 2 feet, 6 inches. So the size of cylinders is: 24½″ + 15½″ × 30″.

EXAMPLE 3.

100 H. P. DELIVERED BY WHEEL, 2 MILES, 5-INCH PIPE.

REQUIRED: Potential at lower end of line.
80 lbs. gauge receiver pressure.

One hundred B. H. P. will compress and deliver 598 cubic feet of free air per minute, or 9.95 cubic feet of free air per second, corresponding to: 1.546 cubic feet per second of cold air at 80 lbs. gauge.

The velocity at entrance in a 5-inch pipe is: 11.35 feet per second, and the absolute pressure at the lower end of the line is $94.7 \times .977 = 92.52$ absolute $= 77.82$ gauge.

Available work:
If air is used cold ....54.6 per cent.
If air is reheated from
    60° to 300° Fahr.. .....79.7 per cent.
If air is reheated from
    60° to 350° Fahr........84.9 per cent.

FUEL CONSUMPTION

Reheating to 300° Fahr ...⅜ cord of wood in 24 hours.
Reheating to 350° Fahr....½ cord of wood in 24 hours.

The total efficiency, taking into account the fuel consumed, is, plain expansion, single cylinder:
    Cold Air................54.6 per cent.
    Reheating to 300°.......75.9 per cent.
    Reheating to 350°....... 79 per cent.

The total efficiency, taking into account the fuel consumed, using compound cylinders and reheating for both high and low pressure cylinders is:
    Reheating to 300°......80.7 per cent.
    Reheating to 350°......83.3 per cent.

EXAMPLE 4.

PUMPING 8 MINER'S INCHES OF WATER 500 FEET HIGH WITH DIRECT-ACTING PUMP.

Consumption of cold free air per minute..........352 cu. ft.
If air is heated (dry) from 60° Fahr. to 300° Fahr.
the consumption falls to ....241 cu. ft.

FUEL CONSUMPTION:

241 cu. ft.=18.412 lbs. air per minute.
1 lb. of air raised in tempera-
    ture by 240° absorbs.....   57.12 B. T. U. per minute.
                        3,427.2 B. T. U. per hour.
                        82,252. B. T. U. per 24 hours.
and 18.412 lbs. will require...1,513,452. B. T. U. per 24 hours.
Assuming 1 lb. wood to yield 5000 B. T. U., and 1 cord= 2000 lbs. the consumption is: .153 or ⅙ cord per 24 hours.

AN EXAMPLE OF A COMPRESSED AIR AND AN ELECTRICAL TRANSMISSION.

To be supplied at the mine:
  100 H. P. to drive motors and
  500 cu. ft. free air per minute, compressed to 80 lbs.
The latter requires 83.5. B. H. P. Now, assuming a motor efficiency of .95 there will be required at motor in the electrical transmission:

$\frac{83.5}{.95}$ = 87.8 H. P 87.8

100 H. P. for machinery.
  $\frac{100}{.95}$ at motor driving machinery............... 105.2
  _____
  Power at lower end of conductor..... 193.0
2 per cent loss on line.
  Power at upper end of line: $\frac{193}{.98}$ =   197
  .95 generator efficiency B.
    H. P. on generator: $\frac{197}{.95}$
  _____
  207

AIR TRANSMISSION.

  500 cu. ft. for drills.
  675 for 100 B. H. P.
  _____
  1175
requiring: 11.75 × 16.7 = 196.22 B H. P.

RIX AIR COMPRESSORS

MANUFACTURED BY THE

Fulton Engineering and Shipbuilding Works
SAN FRANCISCO, CAL.

RIX AIR COMPRESSORS

MANUFACTURED BY THE

Fulton Engineering and Shipbuilding Works
SAN FRANCISCO, CAL.

After the preceding article on the different phenomena and laws, both theoretical and practical, which enter into the subject of compressed air engineering, it seems right and proper to set forth as plainly as possible the different styles and general specifications of the air compressors manufactured especially on this Pacific Coast.

These compressors are all designed and built under the special superintendence of Mr. Edward A. Rix, by the Fulton Engineering and Shipbuilding Works, and are the result of some eighteen years' experience in pneumatics on the Pacific Coast.

There is no doubt that from the conditions under which mining is carried on on the Pacific Coast, one would naturally expect to see a different style and class of air compressor built from those manufactured in the East. The facilities for transportation are vastly different. The special requirement for prospecting plants, which shall be cheap and easily operated, and the tremendous heads of water which are found on the Pacific Coast, necessitate a peculiar construction of compressor, and the large varieties manufactured, descriptions of which follow hereafter, give the intending purchaser or operator ample opportunity to select machines especially fitted to his character of work.

All of the Rix Compressors are of the water-jacket type; that is, the partial cooling during compression is effected by circulating water in a jacket around the cylinder and throughout the heads of the air cylinder. Frequently, also, this circulation is carried within the pistons of the machine, but no water whatever is injected into the cylinder. This method of construction has been constantly followed ever since the manufacture of these machines was begun some eighteen years ago, even though during this time the principal Eastern manufacturers were still enamored of the injection system.

This jacket circulation is not a simple one, and in the smallest of the machines is double, that is, there are two independent water circulations for the machine, the water entering the lower part of the cylinder at two openings, going thence immediately and independently to each head and then around

the body of the cylinder and finally escapi
pendent outlets.

In all cylinders of large diameter or for h
heads are often built with independent circu
manner cold water is assured to many parts o
the same time.

All of the Standard Rix Compressors for o
inlet valves of the poppet type, that is, the va
nuts nor bolts nor threads, and there is nothin
get out of order, and they cannot fall into the
are subjected, of course, to the usual wear a
springs, and these may be taken out in a few
placed as easily.

The outlet valves of the standard machine:
valve type, well known to most all builders c
machinery.

The frames are made in two general style:
liss pattern, and the other of flat bed pattern
head, one being designed for heavy and one f

All of the working parts, such as cranks, l
tons, etc., are made in conformity with the
practise. The crank pins specially are made
so that they do not heat with the intermitter
placed upon them.

The water jackets can be readily cleared o
sediment which may form therein, inasmuch a
are taken off the jackets are completely ex;
very convenient device.

Sight feed lubricators and all necessary oil
fittings are furnished with every machine.

In the tables for the various compressors
no capacities mentioned for cubic feet of
much as the cubic feet of free air will depe:
the piston speed of the machine and inasmu
speed of the machine depends to a great exter
of circumstances, it is deemed easier to use th
to determine the capacities of any of the com;
be noted that the left-hand column conta
diameters of the various sized compressors
the Fulton Engineering Company, and on
column, under the piston speeds mentioned,
various capacities for these cylinders, at th
directly above.

From this table it will be easy to select t
compressor to do the work required, for all t
treatise give the number of cubic feet of air r
various kinds of work. It will only be neces
the total number of cubic feet required and to
speed most advantageously to at once determin
of cylinder. For example, from the requireme
determined that 350 feet of piston velocity
much as is desirable and that the cubic feet

about 550 cubic feet per minute, then an 18½-inch cylinder would be the proper size for a single compressor, or a duplex 14½, making somewhat less than 300 feet of piston velocity per minute.

The question of determining the piston velocity is one of the vital points in the selection of an air compressor. Notwithstanding anything which may be said to the contrary, the most economical compressor is one which moves at a slow piston velocity and high piston velocities are only used to save initial expenditure. Therefore, when one contemplates the installment of a permanent air compressing plant or one which will likely be operated for one or more years, it is always better to select a low piston speed and pay the extra price for the larger machine which this entails, than to pay the extra fuel bill caused by a higher velocity.

It is to be regretted that most purchasers do not understand the value of a low piston speed for an air compressor. A low initial price seems to be the principal virtue. There is not room enough in the ordinary cylinder diameter to give the proper ingress and egress of air under economical conditions. An indicator card from most compressors, running under a piston speed of 400 feet per minute, shows an enormous increase of pressure to force the air through the delivery valves, which, of course means a corresponding loss. The ideal indicator card is one which shows no suction pressure, and which shows that the delivery valves open at, or nearly at, the receiver pressure. Practically, this is not accomplished, and there are few, if any, compressor-builders proud of the indicator card taken from one of their compressing cylinders at such a piston speed. Yet their machines are forced to such speeds, oftener constantly than frequently. The writer has taken cards from various machines that showed 10 per cent of power used in forcing air through the delivery valves. It is not a simple matter to make a practical machine that shall work economically at high piston speed. It is at present far better practise to use a compressor at low piston speed and avoid those losses which cannot be recovered. The ideal system of compression is a continuous one, and while it seems almost impossible, the writer has already built one machine which gives fair promise, and future experiments will probably develop the question. In continuous compression there are no mechanical cylinder losses that amount to much.

No compressor builder advocates high rotative or piston speed, and for the advancement of compressed air practise it is to be hoped that purchasers will consult operative expense rather than initial expenditure.

Diameter of Air Cylinder — Inches	Piston Velocity in Feet per Minute.								
	100	150	200	250	300	350	400	450	500
8	29.7	44.6	59.4	74.3	89.1				
10½	51	76.5	102	127.5	153				
12½	72.3	108.5	144.6	180.8	216.9	253			
14½	97.4	146	195	243.5	292	341	390		
16½	126	189	252	315	378	441	504		
18½	159	239	318	398	477	557	636	878	
20½	195	293	390	488	585	683	780		
22½	235	352	470	587	705	822	940	1058	1175
24½	278	417	556	695	834	973	1112	1251	1390

Cubic Feet of Free Air compressed per Minute in various Air Cylinders at 85% Volumetric Efficiency

G. 37.—For Duplex machines multiply above capacities by 2. For Tandem Duplex machines multiply above capacity by 4.

RIX AIR COMPRESSORS.

Mark or Size	A	B	C	D	E	F
Diameter of Cylinder ins	$2\frac{3}{4}$	$3\frac{1}{4}$	$3\frac{5}{8}$	$2\frac{3}{4}$	$3\frac{1}{8}$	$3\frac{1}{2}$
Length of Stroke Ins.	$6\frac{1}{4}$	$7\frac{1}{4}$	$8\frac{1}{4}$	$6\frac{1}{4}$	$6\frac{1}{2}$	$7\frac{1}{2}$
Strokes per Minute	500	500	500	350	350	350
Cubic Feet of Free Air per Drill and per Minute when the Number of Machines is 1 to 3	64	95	132	52	60	85
3 to 10	57	86	119	47	55	80
10 to 15	51	76	106	42	50	70
15 to 40	45	66	92	37	45	60

Cubic Feet of Free Air at 60° Fahrenheit and at 60 Lbs Gauge Pressure in the Drill Cylinder required per Minute to operate Rix and Giant Rock Drills.

RIX DUPLEX STEAM ACTUATED COMPRESSOR.

CLASS A, FIG. 32.

Fig. 32 is a half-tone of the Rix Duplex Steam Actuated Compressor, of the flat bed type, having slipper cross head.

Fig. 33 is a plan of this same machine, showing arrangements of foundation bolts and piping.

Fig. 34 is an end elevation of the same compressor.

Fig. 35 is a side elevation of the same compressor, and is at the same time a side elevation of the Single steam actuated compressor.

Wherever possible, it is desirable to install a Duplex Air Compressor. The cranks being placed at right angles, the air is discharged more continuously throughout the whole revolution, and the result is that the strains in the machine are more evenly divided and the machine as a whole gives better satisfaction.

Another reason which should prompt a Duplex machine is, that should it be necessary to discontinue the use of one-half of the machine for repairs, the other half is always available and is a complete working compressor in itself.

The following is a table of dimensions for these compressors:

RIX DUPLEX STEAM ACTUATED COMPRESSOR.

CLASS A.

For Revolutions per minute, Cubic Feet Free Air, Rock Drill Capacity, see pages 84 and 85.

No.	Diameter Steam Cylinder.	Diameter Air Cylinder.	Stroke.	H. P. Boiler.	Price.
1	10	10½	14	60	......
2	12	12½	16	80	......
3	14	14½	18	110	....
4	16	16½	18	140	......
5	18	18½	24	200	....
6	20	20½	24	2 0	....
7	22	22½	30	310	....
8	24	24½	30	400	....

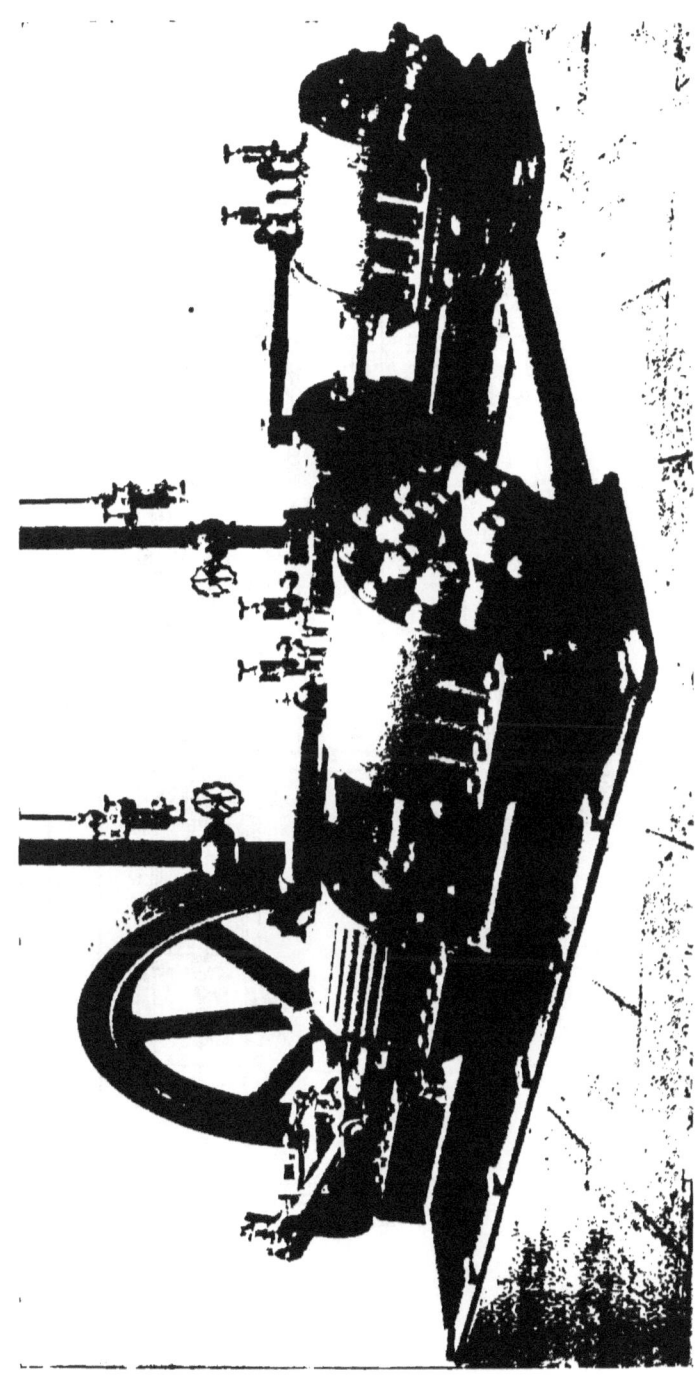

FIG. 32—Class A.—Rix Duplex Steam Actuated Compressor. Mfd. by Fulton Engineering and Shipbuilding Works, San Francisco.

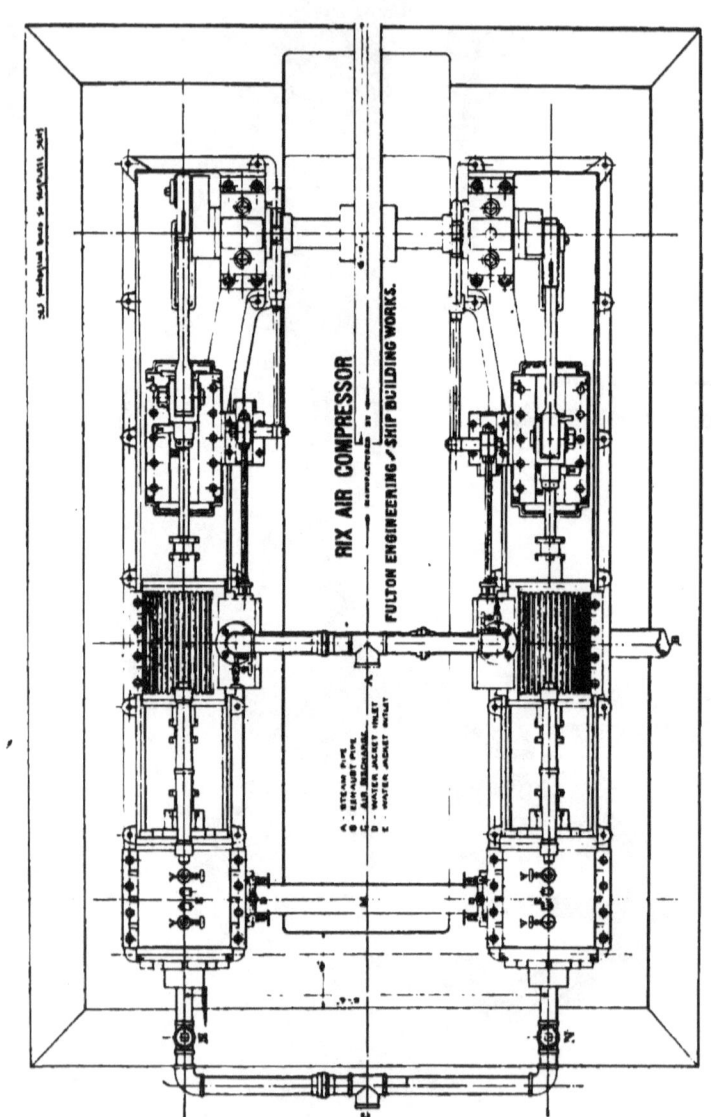

FIG. 33.—Class A.—Rix Duplex Steam Actuated Compressor.

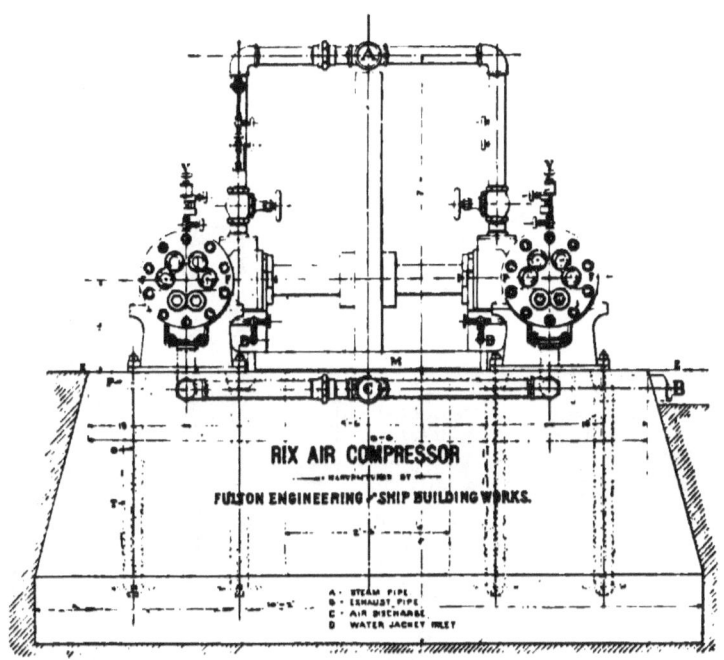

FIG. 34—Class A.—Rix Duplex Steam Actuated Compressor.

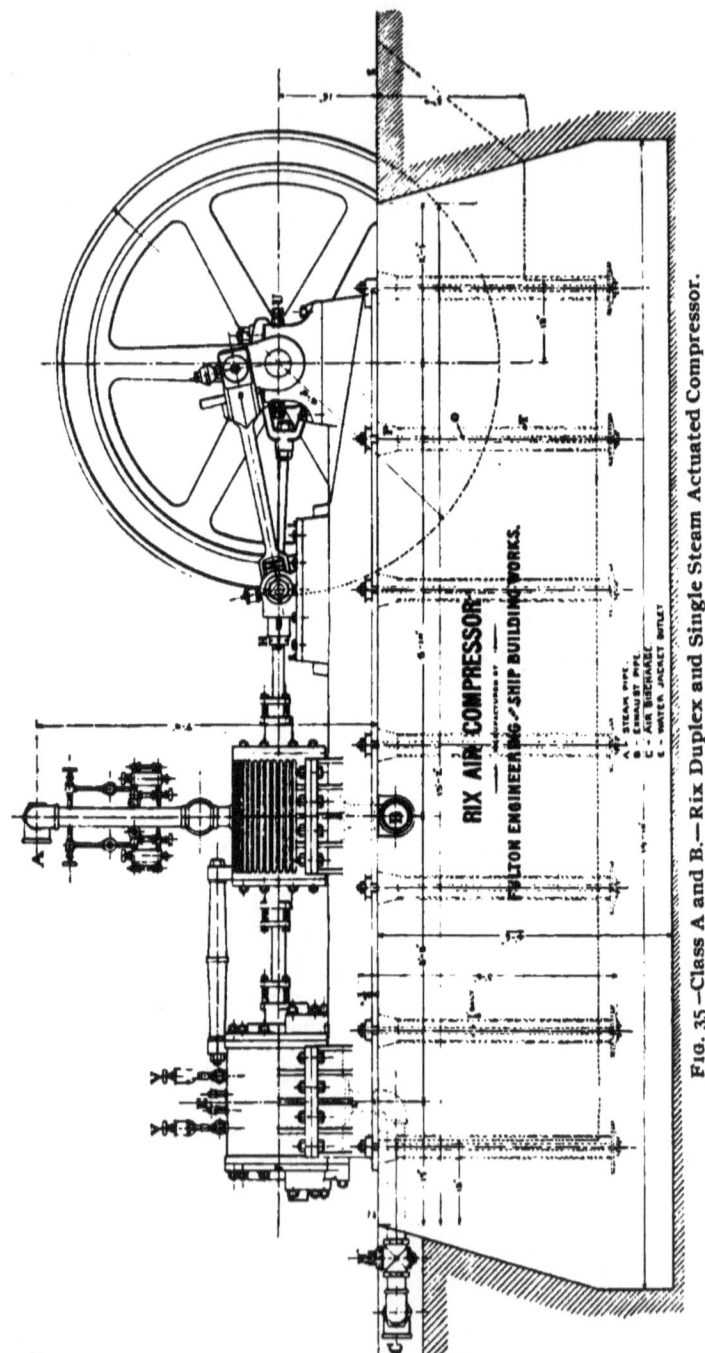

Fig. 35.—Class A and B.—Rix Duplex and Single Steam Actuated Compressor.

RIX SINGLE STEAM ACTUATED COMPRESSOR.

CLASS B, FIG. 36.

The following half-tone, Fig. 36, shows the general style of construction of Class B, Rix Single Steam Actuated Compressor, and Fig. 35 shows the side elevation of same.

This machine differs only from the Duplex Compressor in the fact that it is one-half of that machine and has an outboard bearing.

The following is a table of the various and proper dimensions.

RIX SINGLE STEAM ACTUATED COMPRESSOR.

CLASS B.

For Revolutions per minute, Cubic Feet Free Air, and Rock Drill Capacity, see pages 84 and 85.

No.	Diameter. Steam Cylinder	Diameter Air Cylinder.	Stroke.	H P. Boiler	Price.
9	10	10½	14	30	....
10	12	12½	16	40	....
11	14	14½	18	55	....
12	16	16½	18	70	....
13	18	18½	24	100	....
14	20	20½	24	130	....
15	22	22½	30	155	....
16	24	24½	30	200	....

FIG. 36.—Class B.—Rix Single Steam Actuated Compressor. Mfd. by Fulton Eng. and Shipbuilding Works, San Francisco.

RIX SINGLE STEAM ACTUATED COMPRESSOR, SELF-CONTAINED TYPE.

CLASS C, FIG. 38.

This machine is one which is offered to the mining public as the least expensive and most generally useful machine of the kind ever constructed. It will be noted from the half tone that this consists of an independent standard engine on a bed-plate connected to an air-compressing cylinder, the whole being tied together for proper operation. The engine is self contained, there being no outboard box, the fly wheel pulley being overhung, so that this machine can be placed anywhere and is ready for operation at once. A belt can be placed upon the fly wheel pulley and be used to operate a pump or any other machine that may be desired while the compressor is not in use, in which case it will only be necessary to remove one inlet valve on each end of the air cylinder and the compressor end of the machine becomes inactive.

This machine is especially built for prospecting, temporary work and for experiments, where a permanent plant is too expensive. It will be noted, from the construction, that the engine can be entirely removed and used independently should occasion demand, and the whole arrangement is one which gives a prospector an opportunity to easily dispose of his machine should his mining venture prove a poor one.

The following is the list of sizes of the Class C Compressor.

RIX SINGLE STEAM ACTUATED COMPRESSOR, SELF-CONTAINED.

CLASS C.

For Revolutions per minute, Cubic Feet Free Air, and Rock Drill Capacity, see pages 84 and 85.

No.	Diameter Steam Cylinder	Diameter Air Cylinder	Stroke.	H. P. Boiler	Price
17	7	8	10	15	....
18	8	8	10	20	....
19	9	10½	12	25	....
20	10	10½	12	30	....
21	10	11½	14	30	....
22	11	11½	14	35	....
23	12	12½	16	40	....
24	13	12½	16	45	....
25	14	14½	18	55	....
26	16	16½	20	70	....
27	18	18½	22	100	....

Fig. 38.—Class C.—Rix Single Steam Actuated Compressor, Self Contained.

RIX DUPLEX SHAFT-DRIVEN COMPRESSOR.

CLASS D, FIG. 39.

This half-tone represents one of the new style Shaft-Driven Rix Duplex Compressors, heavy duty style. This machine has Corliss frame, extra large wrist pins, and large cross head. The frame is swelled up on the front head so that the head may be removed without disconnecting the cylinder. The Compressor which was the subject for this half-tone was driven by a twelve-foot tangential water wheel, under a head of two hundred and seventy-five feet. It may, however, be driven by belt.

Fig. 40 is a side elevation of the same machine, showing belt pulley.

Fig. 41 shows a sectional machine of the same class, but having a flat bed, with water wheel attached upon the shaft.

This Compressor, as all the sectional compressors hereinafter mentioned, is made in sections not to exceed 325 lbs. in weight, so that they may be carried upon mules.

The following table gives the sizes and principal dimensions for the Class D machines:

RIX DUPLEX SHAFT-DRIVEN COMPRESSORS.
CLASS D.

For Revolutions per minute, Cubic Feet Free Air, and Rock Drill Capacity, see pages 84 and 85.

No.	Diameter Air Cylinder.	Stroke.	Price.
28	8	12	....
29	10½	14	....
30	13	16	....
31	14½	18	....
32	16½	18	....
33	18½	24	....
34	20½	24	....
35	22½	30	....
36	24½	30	....

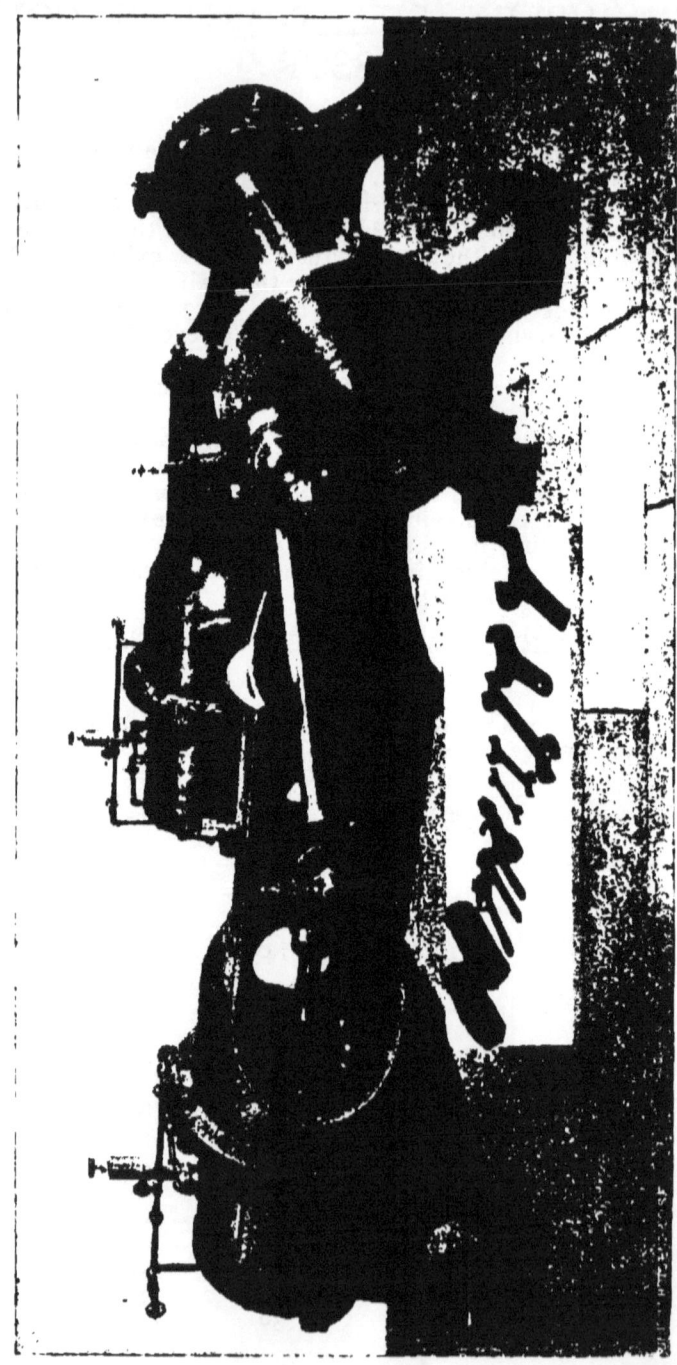

Fig. 39.—Class D.—Rix Duplex Shaft-driven Compressor. Mfd. by Fulton Engineering and Shipbuilding Works, San Francisco.

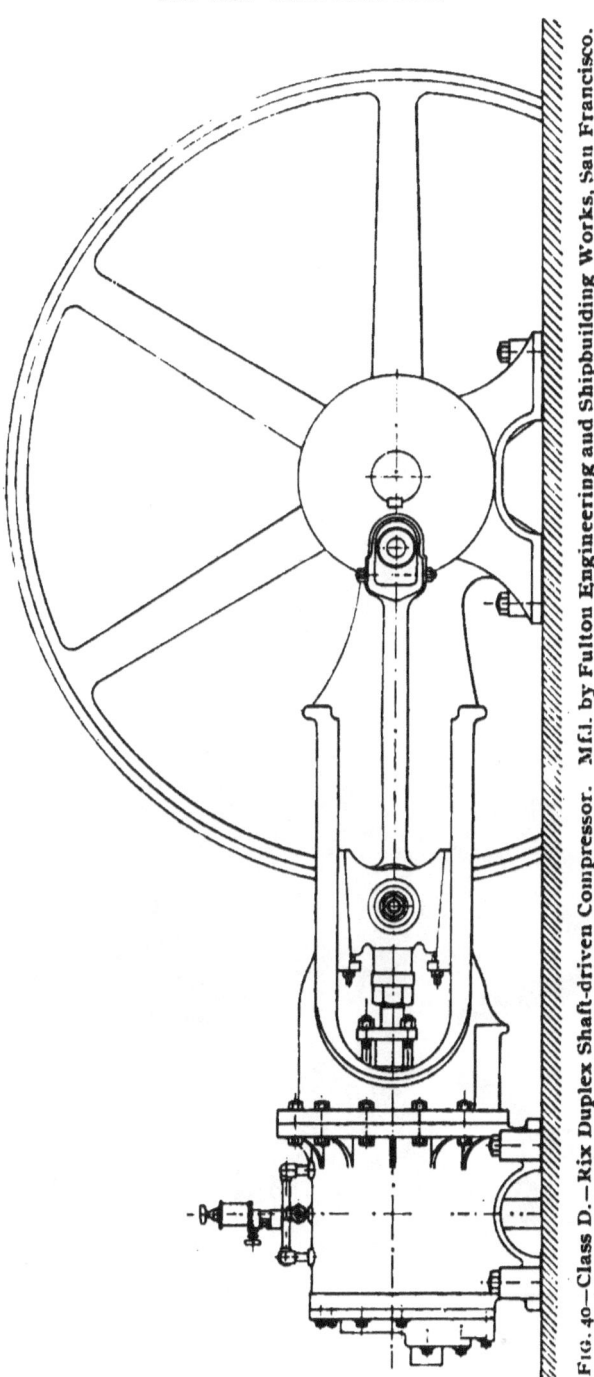

FIG. 40—Class D.—Rix Duplex Shaft-driven Compressor. Mfd. by Fulton Engineering and Shipbuilding Works, San Francisco.

FIG. 41.—Class D.—Rix Duplex Sectional Shaft-driven Compressor. Manufactured by Fulton Engineering and Shipbuilding Works, San Francisco.

RIX DUPLEX TANDEM SECTIONAL SHAFT-DRIVEN COMPRESSORS.

CLASS E, FIG. 42.

These Compressors are entirely similar to the Class D Machines as noted in Fig. 41, with the exception that the bed is extended and an additional air cylinder placed tandem to the others. This makes a very convenient form of machine, and one which gives a large air capacity with little additional weight. These air cylinders are so connected up that any one of the four cylinders, or any combination of the four cylinders, may be run together. The utility of this machine will be recognized at once.

Fig. 43 is a side elevation of this Class E machine.

RIX DUPLEX TANDEM SECTIONAL SHAFT-DRIVEN COMPRESSORS.

CLASS E.

For Revolutions per minute, Cubic Feet Free Air, and Rock Drill Capacity, see pages 84 and 85.

No.	Diameter Air Cylinder.	No. of Air Cylinders.	Stroke.	Price.
37	8	4	12	....
38	10½	4	14	....
39	12½	4	16	....
40	14½	4	18	....

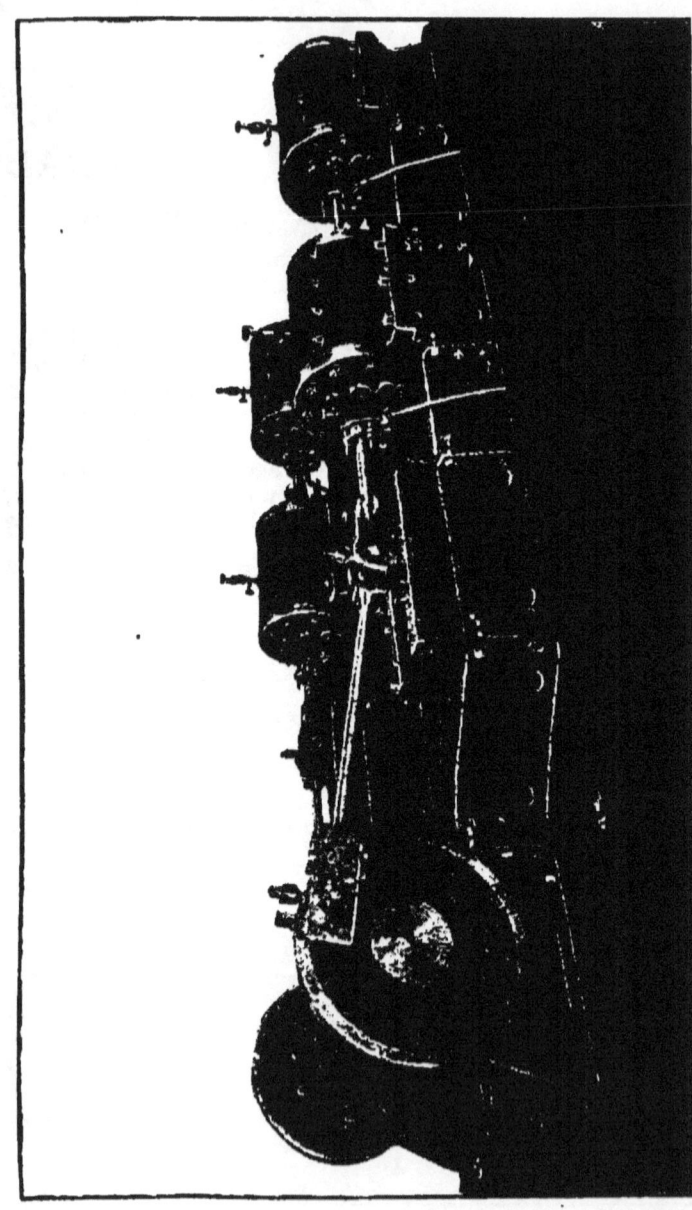

F G. 42.—Class E.—Rix Duplex Tandem Sectional Shaft-Driven Compresor. Manufactured by Fulton Engineering and Shipbuilding Works, San Francisco.

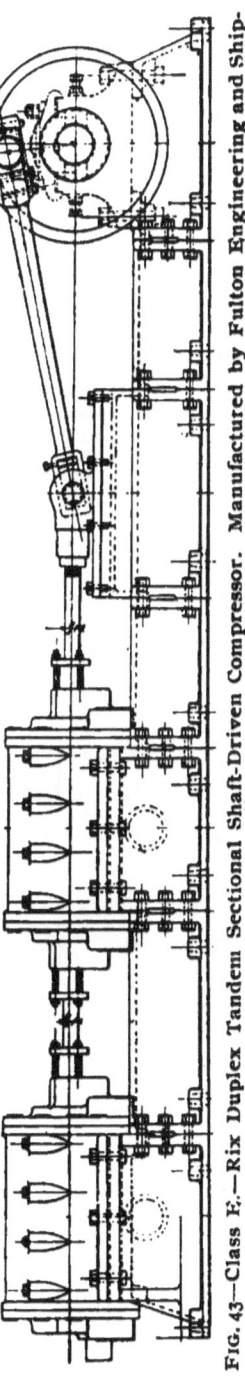

Fig. 43.—Class E.—Rix Duplex Tandem Sectional Shaft-Driven Compressor. Manufactured by Fulton Engineering and Shipbuilding Works, San Francisco.

RIX SINGLE SHAFT-DRIVEN COMPRESSOR.

Class F, Fig. 44.

This half-tone shows a flat bed type of compressor, but they are made also with Corliss frames, as shown in the Class D machines, Fig. 40, the smaller machines being made as per Fig. 44. This machine has an outboard bearing and may be driven either by belt, pulley, or by water wheel upon the shaft.

Fig. 45 shows a side elevation of this Class F compressor.

The following is a table of the sizes and general dimensions of this style of air compressor:

RIX SINGLE SHAFT-DRIVEN COMPRESSOR.

CLASS F.

For Revolutions per minute, Cubic Feet Free Air, and Rock Drill capacity, see pages 84 and 85.

No.	Diameter Air Cylinder.	Stroke.	Price.
41	8	12	....
42	10½	14	....
43	13	16	....
44	14½	18	....
45	16½	18	....
46	18½	24	....
47	20½	24	....
48	22½	30	....
49	24½	30	....

FIG. 44—Class F.—Rix Single Shaft-driven Compressor.

104 RIX AIR COMPRESSOR

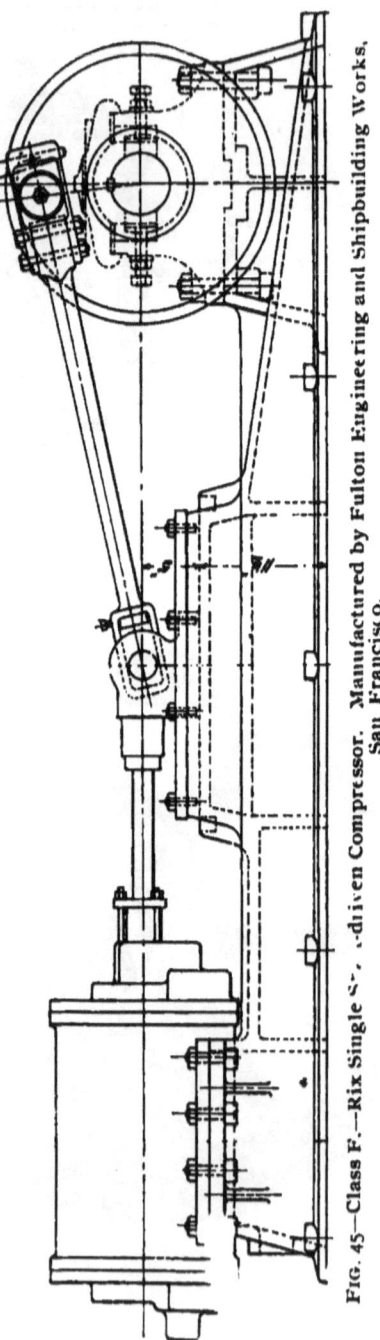

FIG. 45.—Class F.—Rix Single Steam-driven Compressor. Manufactured by Fulton Engineering and Shipbuilding Works, San Francisco.

RIX COMBINED DUPLEX STEAM ACTUATED AND SHAFT-DRIVEN COMPRESSOR.

CLASS G, FIG. 46-46½.

This is a form of compressor which is especially adapted to the wants of the Pacific Coast, where there is abundance of water supply during one portion of the season and an insufficient supply during the remainder. It becomes, therefore, necessary to run the compressor with water power during a portion of the year, and steam power during the balance.

It will be noted from the half tone that the air cylinders are placed next to the water wheel, which water wheel has been built upon the fly wheel of the machine, the steam cylinders being tandem to the air cylinders, with a sleeve coupling between. When it is desired to run by water power it is only necessary to remove the sleeve coupling, and the machine becomes a water power compressor. The couplings may be replaced in an hour, at any time, and the machine again converted into a duplex steam machine, using the combined fly wheel and water wheel for a fly wheel.

These compressors are made in the following sizes:

RIX COMBINED DUPLEX STEAM ACTUATED AND SHAFT-DRIVEN COMPRESSOR.

CLASS G.

For Revolutions per minute, Cubic Feet Free Air, and Rock Drill Capacity, see pages 84 and 85.

No.	Diameter Steam Cylinder	Diameter Air Cylinder	Stroke	H. P. Boiler	Price
50	10	10½	14	60	....
51	12	12½	16	80	....
52	14	14½	18	110	....
53	16	16½	18	140	....
54	18	18½	24	200	....
55	20	20½	24	260	....
56	22	22½	30	310	....
57	24	24½	30	400	....

FIG. 46—Class G.—Rix Combined Duplex Steam Actuated and Shaft-driven Compressor. Manufactured by Fulton Engineering and Shipbuilding Works, San Francisco.

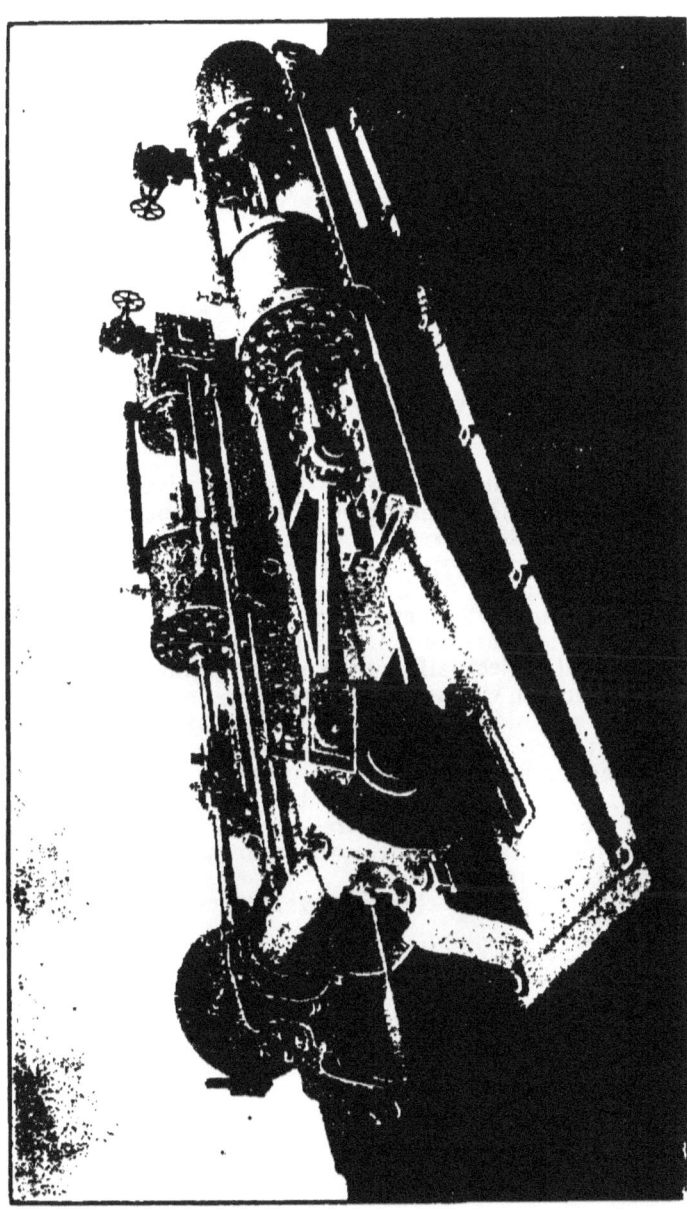

FIG. 46½.—Class G.—Rix Combined Steam Actuated Shaft-driven Compressor. Manufactured by Fulton Engineering and Shipbuilding Works, San Francisco.

RIX STEAM ACTUATED VERTICAL COMPRESSORS.

CLASS H, FIG. 47.

This style of compressor is one which has given universal satisfaction in this State, a machine of similar type having run continuously from 1880 to the present date with no expense whatever beyond valve springs. It is single acting; the air cranks being placed at 180 degrees from each other, which balances the machine completely, and the cylinders being vertical there is no internal wear of any consequence. The steam engine is placed horizontally on the floor, for the double purpose of keeping the warmth of the steam cylinder away from the inlet air, and also for the purpose of making the steam crank at right angles to the air cranks.

This compressor is made in only one size: 12-inch steam cylinders, 12½-inch air cylinders by 16-inch stroke, and catalogued No. 58. Capacity in free air per minute, see page 84, both cylinders being the equivalent of one double-acting 12½-inch cylinder, as per table.

Figs. 48, 49, and 50 show different views of this same machine.

FIG. 47—Class II.—Rix Steam Actuated Vertical Compressor. Manufactured by Fulton Engineering and Shipbuilding Works, San Francisco.

FIG. 48—Class H.—Rix Steam Actuated Vertical Compressor. Manufactured Fulton Engineering and Shipbuilding Works, San Francisco.

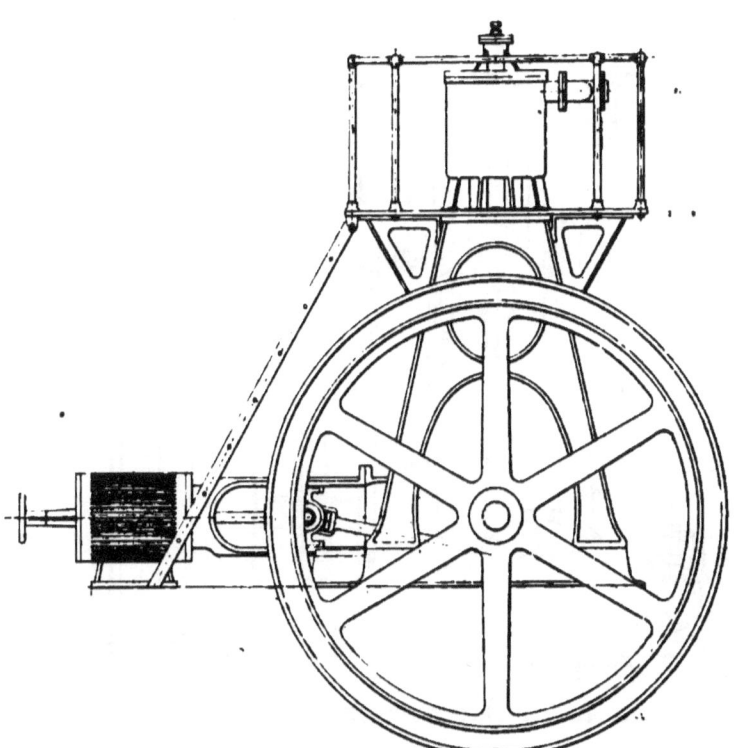

FIG. 49—Class H.—Rix Steam Actuated Vertical Compressor. Manufactured by Fulton Engineering and Shipbuilding Works, San Francisco.

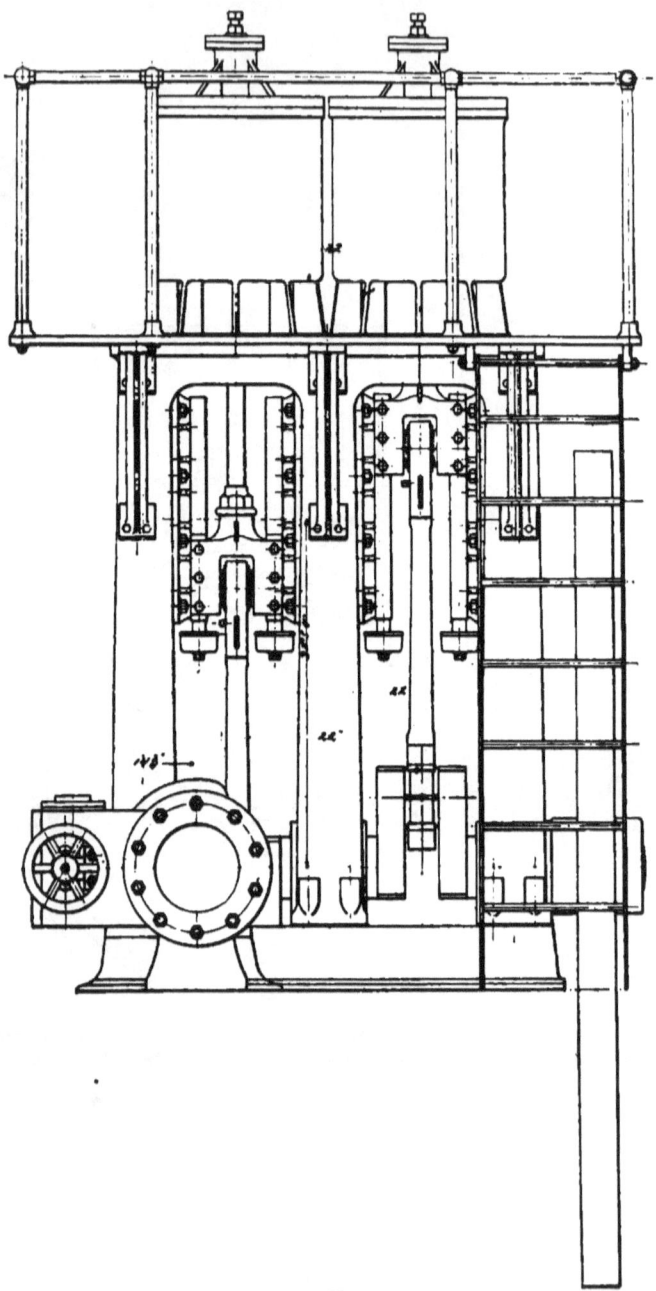

FIG. 50—Class H.—Rix Steam Actuated Vertical Compressor. Manufactured by Fulton Engineering and Shipbuilding Works, San Francisco.

RIX SINGLE CORLISS ACTUATED COMPRESSORS.

CLASS I, FIG 51.

These Compressors consist of a Standard Corliss engine, to which there is placed tandem the air cylinder.

Fig. 52 shows a plan of the single machine. They are an economical and high-class machine in every respect.

The following table shows the sizes and dimensions of the Class I, Rix Single Corliss Actuated Compressors:

RIX SINGLE CORLISS ACTUATED COMPRESSORS.

CLASS I.

For Revolutions per minute, Capacity Free Air, Rock Drill Capacity, see pages 84 and 85.

No.	Diameter St'm Cylinder.	Diameter Air Cylinder.	Stroke.	Price.
59	12	12½	30	....
60	12	14½	30	....
61	14	14½	30	....
62	14	16½	30	....
63	16	16½	30	....
64	16	18½	30	....
65	16	16½	36	....
66	16	18½	36	....
67	16	16½	42	....
68	16	18½	42	....
69	18	18½	36	....
70	18	20½	36	....
71	18	18½	42	....
72	18	20½	42	....
73	18	18½	48	....
74	18	20½	48	....

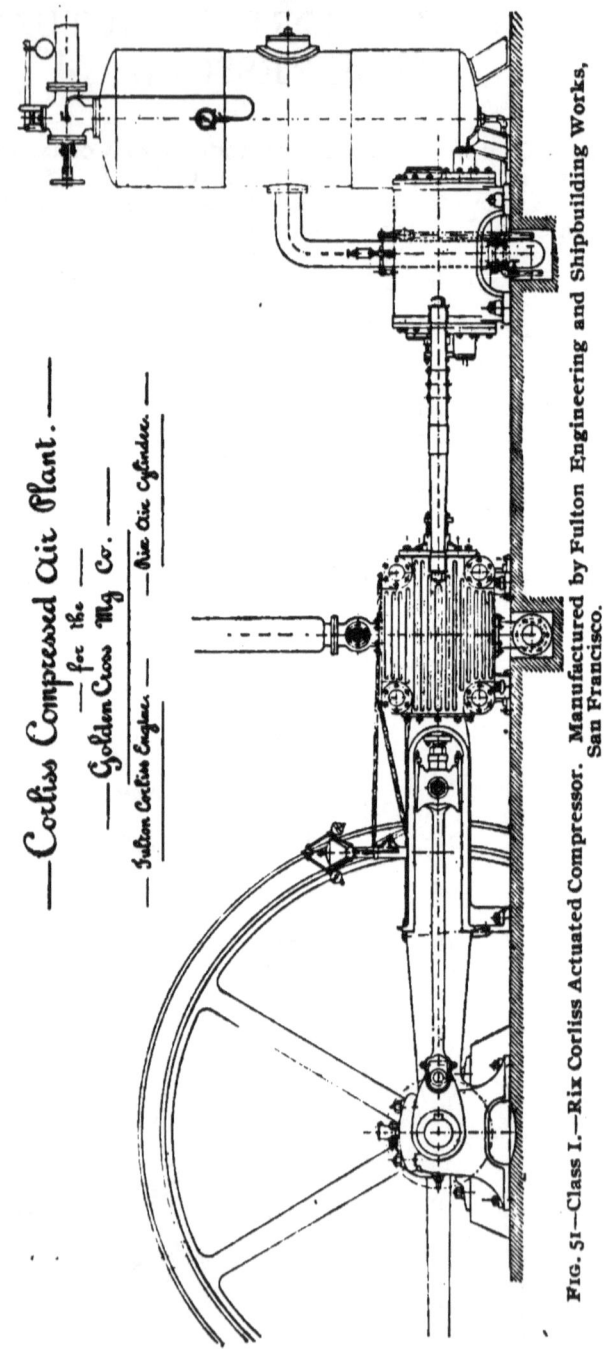

FIG. 51.—Class I.—Rix Corliss Actuated Compressor. Manufactured by Fulton Engineering and Shipbuilding Works, San Francisco.

RIX AIR COMPRESSORS.

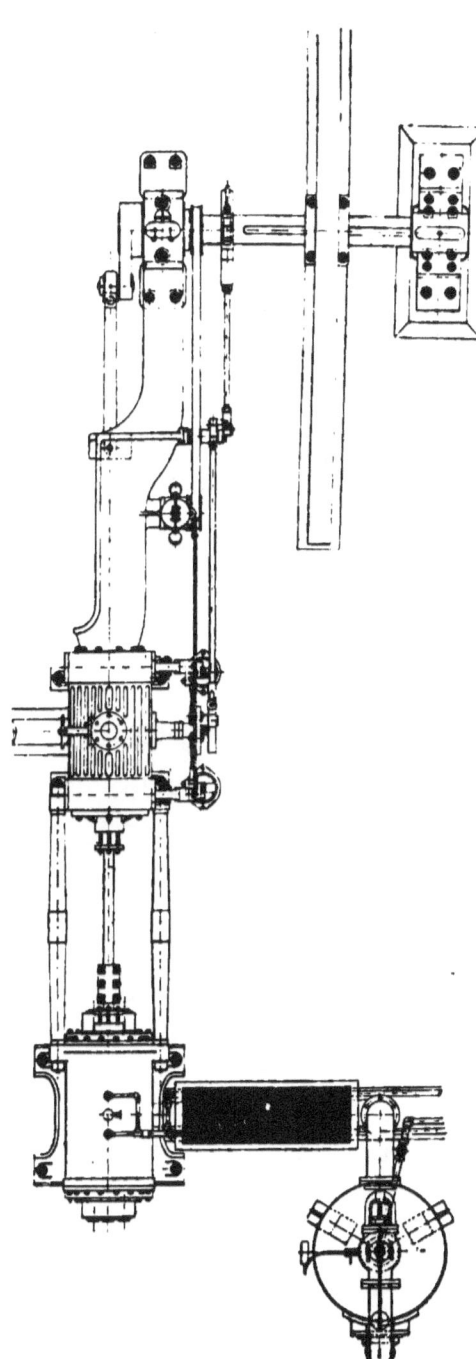

FIG. 52—Class I.—Rix Corliss Actuated Compressor. Manufactured by Fulton Engineering and Shipbuilding Works, San Francisco.

RIX COMPOUND CORLISS ACTUATED COMPRESSORS.

Class J comprises the *Rix Compound Corliss Actuated Compressors*, which are entirely similar to those of Class I. excepting that the steam cylinders are compound, the air cylinders being alike.

The following is a table showing the sizes and principal dimensions of the Class J Compressors:

RIX COMPOUND CORLISS ACTUATED COMPRESSORS.
CLASS J.

For Revolutions per minute, Cubic Feet Free Air, Rock Drill Capacity, see pages 84 and 85.

No.	Diameter High Pressure.	Diameter Low Pressure.	Diameter Air Cylinder.	Stroke.	Price.
75	12	22	12½	30	....
76	12	22	14½	30	....
77	14	26	14½	30	....
78	14	26	16½	30	....
79	16	30	16½	30	....
80	16	30	18½	30	....
81	16	30	16½	36	....
82	16	30	18½	36	....
83	16	30	16½	42	....
84	16	30	18½	42	....
85	18	34	18½	36	....
86	18	34	20½	36	....
87	18	34	18½	42	....
88	18	34	20½	42	....
89	18	34	18½	48	....
90	18	34	20½	48	....

Both the Compressors Class I or Class J are furnished either condensing or non-condensing.

RIX LIGHT DUTY COMPRESSOR OR VACUUM PUMP.

CLASS K, FIG 53.

This Compressor is adapted for very light work and is a self-contained machine working from a Scotch yoke. It is intended for pressures up to 25 lbs. only, and can be either used as a compressor or a vacuum pump, the valves being arranged for that purpose. It is single acting and the discharge is absolutely complete, there being no clearance whatever. It is capable of creating a 29-inch vacuum.

Made in four sizes having 4", 5", 6", and 7" diameter of cylinders, and catalogued No. 91, 92, 93, and 94 respectively.

This machine is a very inexpensive and satisfactory compressor to have in laboratories, shops, and canneries, or for blowing crude oil into furnaces. A four-inch belt is ample to run any of them. The peculiar feature which is advantageous as a vacuum pump is the discharge valve which covers the whole end of the cylinder. The piston touches it, moves it slightly from its seat, thus dispelling all the air, the valve reseating as the piston begins the return stroke.

FIG. 53—Class K.—Rix Light Duty Compressor or Vacuum Pump.

RIX STEAM ACTUATED DUPLEX COMPRESSORS.

CLASS L.

These compressors are designed for compressing air to not exceeding twenty-five pounds per square inch, with a steam pressure at from sixty to ninety pounds. They are made with Scotch Yoke, as may be seen from the cut in Fig. 54, and are self contained in every respect. They are especially adapted for this Coast, for furnishing air for burning crude petroleum or distillate.

These machines are far heavier and stronger than any machine which is built in the East for the same purpose; the same comparative cylinder sizes being made about twenty-five per cent heavier, so that for use on shipboard they may be absolutely relied upon not to break or give out when at service.

These machines are complete with all lubricators, valves, and also automatic governor, which will regulate the machine to within two or three pounds of the receiver pressure.

Each one of these compressors is set up in the shop and thoroughly tested before shipment, so that the machine will be ready to go to work as soon as set upon its foundations.

The following are the sizes of the *Rix Steam Actuated Duplex Compressors, Class L:*

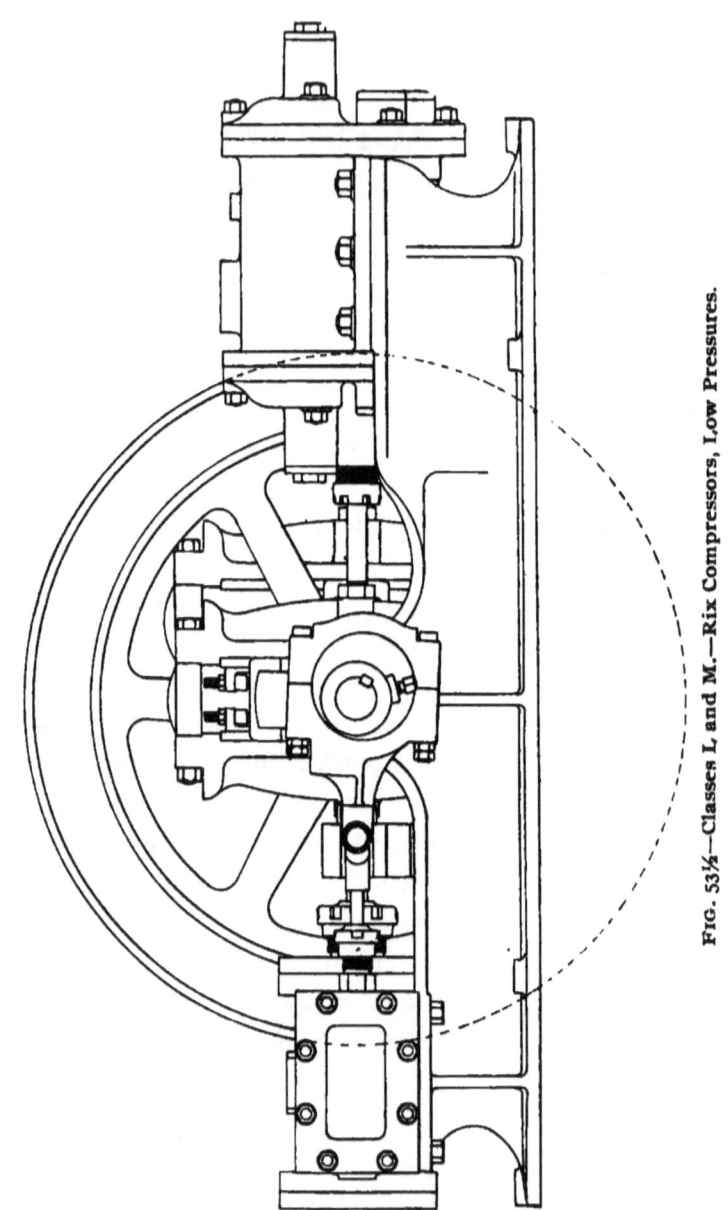

FIG. 53½.—Classes L and M.—Rix Compressors, Low Pressures.

RIX STEAM ACTUATED SINGLE AIR COMPRESSORS.

CLASS M.

These machines are precisely like those of Class L, excepting that they are Single instead of Duplex, and are fitted up in precisely the same manner.

They are complete with governor, lubricators, oilers, and wipers.

Each machine is tested before leaving the shop, so that it is ready for work immediately it is erected upon its foundations.

The following are the sizes of the *Rix Steam Actuated Single Air Compressors, Class M:*

RIX STEAM ACTUATED DUPLEX COMPRESSORS, CLASS L.

No.	Diameter Steam Cylinder in Inches.	Diameter Air Cylinder in Inches.	Length of Stroke in Inches.	Steam Supply in Inches.	Steam Exhaust in Inches.	Air Inlet in Inches.	Air Discharge in Inches.	Revolutions per Minute.	Cubic Feet Free Air per Minute.	Horse Power Required.	Price.
95	4	6	6	1	1¼	1¼	1½	130	51	5	
96	5	7	7	1¼	1½	1½	2	120	74	7	
97	5	8	7	1¼	1½	2	2½	120	98	7	
98	6	9	9	1½	1½	2	2½	120	159	10	
99	7	10	9	1½	2	2	2½	120	196	13	
100	9	14	9	2	2½	3	3½	120	384	20	

RIX STEAM ACTUATED SINGLE AIR COMPRESSORS, CLASS M.

Number.	Diameter Steam Cylinder in Inches.	Diameter Air Cylinder in Inches.	Length of Stroke in Inches.	Steam Supply in Inches.	Steam Exhaust in Inches.	Air Inlet in Inches.	Air Discharge in Inches.	Revolutions per Minute.	Cubic Feet of Free Air per Minute.	Horse Power Required.	Price.
101	4	6	6	¾	1	1¼	1¼	130	25	2½	
102	5	7	7	1	1	1½	1½	120	37	3½	
103	5	8	7	1	1	2	2	120	49	3½	
104	6	9	9	1	1¼	2	2	120	79	5	
105	7	10	9	1¼	1½	2	2	120	98	7	
106	9	14	9	1½	2	3	3	120	192	10	

CLASS N, FIG. 54—Duplex Direct Acting Steam Actuated Compressors.

DUPLEX DIRECT ACTING STEAM ACTUATED COMPRESSORS.

CLASS N.

It will be noted from the cut, Figure 54, that these compressors are made after the style of the DIRECT ACTING STEAM PUMP, and they are designed to meet certain requirements where light pressures and inexpensive or temporary machinery are desired. They are the least expensive of all compressors which are built, and while they do not have a very high volumetric efficiency, they are easily installed and for certain classes of work are amply economical.

The AIR CYLINDERS are composition lined and the PISTON rods are of brass. Every machine is fitted complete with its PROPER LUBRICATOR and wrenches. The VALVE MECHANISM is so arranged that the air pistons work against a constant pressure at all times, thus obtaining quite a high efficiency for this character of compressor, and insuring a uniform stroke.

There are no DEAD CENTERS on the machine, and the pump is consequently always ready to start. The dispensing of the crank and flywheel renders it possible to place this compressor in an extremely small space.

The VALVES in the steam end are slide valves, and in the air and poppet valves of the ordinary type positively controlled by the valve mechanism. The entire apparatus is compact, durable, and self-contained. There are no intricate working parts whatever, and it requires very little attention to operate it.

As a general rule it is desirable to operate this machine in connection with a PRESSURE REGULATOR, which we furnish with the machine if desired. The PRESSURE REGULATOR automatically controls the speed, slowing down and finally stopping the pump when the desired air pressure is obtained, and gradually starting up again when the air is exhausted from the reservoir. This regulator practically makes the machine automatic in its operation.

This Compressor is used in BREWERIES for BEER RACKING, and is especially desirable for that purpose. It is also used in running PNEUMATIC TOOLS for cutting marble or granite, or other building stone, and also for CHIPPING and CALKING BOILERS; for the running of SAND BLASTS; for the handling of ACIDS in refineries; for running small PNEUMATIC CRANES; for use in RUBBER FACTORIES, or for pumping pressures upon AUTOMATIC FIRE EXTINGUISHERS; for CLEANING CARS where a jet of air is used to dust off cushions it is especially valuable as an inexpensive and cheap machine; for the running of CLIPPING MACHINES, or for running COAL CONVEYORS, or SMALL ROCK DRILLS, where pressures not exceeding fifty or sixty pounds are required; for PNEUMATIC EJECTORS, or for producing vacuums for FILTERING purposes; and the enumerable requirements where low pressure compressed air is desired.

For RUNNING ROCK DRILLS we do not advocate it for a permanent plant, but for a prospecting plant for small drills, these compressors can be readily installed and will prove first-class in their operation.

These Compressors are particularly adapted for furnishing the compressed air to BURN PETROLEUM COMPOUNDS UNDER BOILERS FOR GENERATING STEAM.

DUPLEX DIRECT ACTING COMPRESSORS.

Class N.

Capacities calculated on piston speed of 60 feet and volumetric efficiency of 70 per cent.

No.	Diameter of Steam Cylinder.	Diameter of Air Cylinder.	Stroke.	Cubic Feet of Free Air.	Size Steam Pipe.	Size Air Pipe.	Maximum Air Pressure.	Price.
107	4½	3	4	2.07	½	¾	60	......
108	4½	4	4	3.68	½	¾	50	......
109	4½	4½	4	4.65	½	¾	40	......
110	5¼	3	5	2.07	¾	1	60	......
111	5¼	3½	5	2.81	¾	1	50	......
112	5¼	4	5	3.68	¾	1	50	......
113	5¼	4½	5	4.65	¾	1	45	......
114	5¼	4¾	5	5.06	¾	1	40	......
115	5¼	6	5	8.28	¾	1	20	......
116	6	3	6	2.07	1	1¼	70	......
117	6	3½	6	2.81	1	1¼	60	......
118	6	4	6	3.68	1	1¼	55	......
119	6	4½	6	4.65	1	1¼	50	......
120	6	4¾	6	5.06	1	1¼	45	......
121	6	6	6	8.28	1	1¼	40	......
122	6	6½	6	9.70	1	1¼	30	......
123	6	7	6	11.27	1	1¼	25	......
124	6	7½	6	13.	1	1¼	20	......
125	6	8	6	14.72	1	1¼	15	......

PNEUMATIC GOVERNORS.

Fig. 54½ shows the *Pneumatic Governor* which the Fulton Engineering Company attach to all the Corliss Compressors. This Governor consists in a special attachment arranged in connection with the Standard Corliss Governor, which is actuated by the air pressure. When the pressure rises in the air receiver the Governor balls are automatically lifted and the hooks are thus tripped independently of the number of revolutions which the engine is making. When the pressure falls in the tank the device drops out of the way and the engine is controlled by the Corliss Governor pure and simple.

For all ordinary compressors, when desired, a Governor is furnished which controls the admission of steam readily as the load varies. It is simple and effective in its operation.

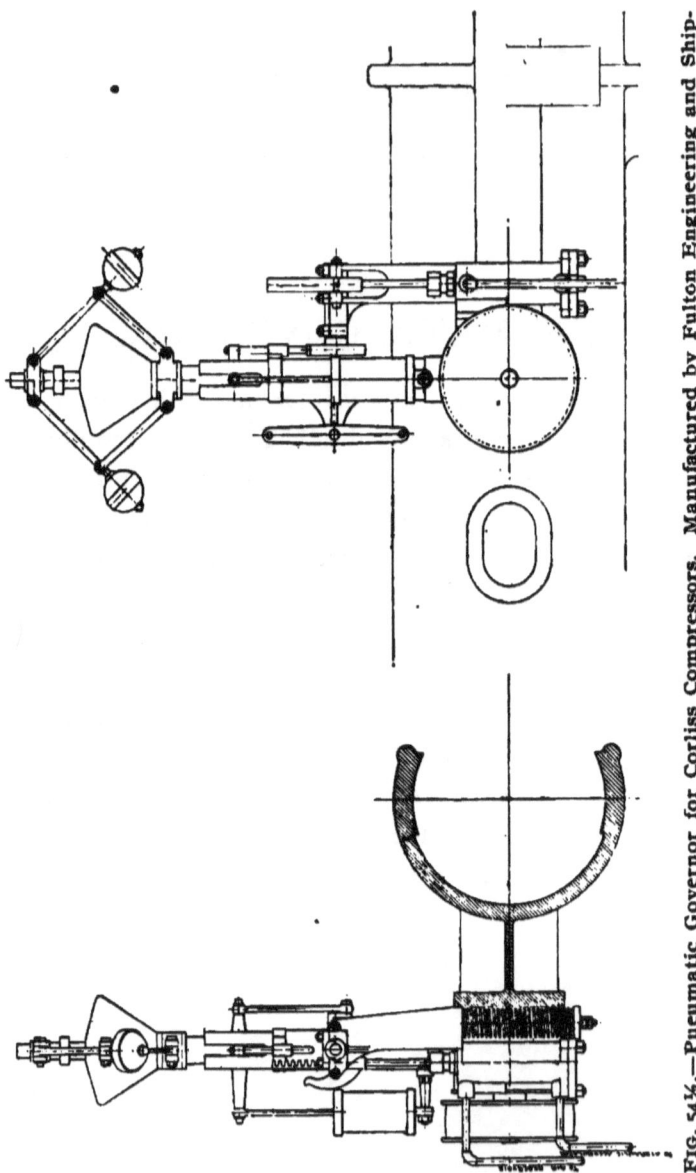

Fig. 54½.—Pneumatic Governor for Corliss Compressors. Manufactured by Fulton Engineering and Shipbuilding Works, San Francisco.

THE RIX COMPOUND COMPRESSOR.

In speaking of the various means in practise for cooling the air during its compression, reference has been made heretofore in this treatise to compounding the compressing cylinders. The advantages of this process are so important that it has come into general use and Compound Compressors nowadays are beginning to be the rule rather than the exception. It is therefore interesting to give some explanation of this method of compression.

The principle of Compound Compression can be described as follows: Suppose that a certain volume of air at atmospheric pressure and temperature is to be raised to a certain pressure and delivered into a receiver; in ordinary, or single stage compression, this air is introduced into a cylinder wherein a piston effects the compression and delivery of that air at each stroke. This compression, as we know, and especially in fast moving machines, is accompanied by a considerable development of heat, which causes a loss of efficiency.

In the compound machine, air is admitted into a cylinder, as before, but it is compressed and delivered into a receiver at a pressure smaller than the desired final pressure. In this first period or stage of compression there is a certain amount of heat developed, less, however, than in the single stage machine. The compressed air, after it is delivered into this first receiver at the intermediate pressure, is cooled by coming in contact with a number of copper tubes through which cold water is rapidly circulated. This receiver is quite similar to the surface condenser used in marine engines and is termed the Intercooler, and the compressed air leaves it after having been deprived of its heat, and reduced to practically the temperature of the water. It is then admitted into another smaller cylinder wherein its pressure is raised by another piston—the air being again passed through another intercooler—then admitted into a third cylinder, and so on until the final desired pressure is reached.

The compression of air, instead of being affected all at once, is therefore performed in several stages, each separated from the following one by a cooling to the atmospheric temperature. It may be readily conceived that the partial amounts of heat developed in this series of cylinders are more effectively dealt with than when the whole amount of heat is liberated in a single cylinder. On this ground the Compound Compressor will therefore possess a higher efficiency than the single stage machine.

Another advantage is that the variation of load on the piston during the stroke is less in the compound, and consequently the strains on the crankpins are reduced, and a lighter

130   THE RIX COMPOUND COMPRESSOR.

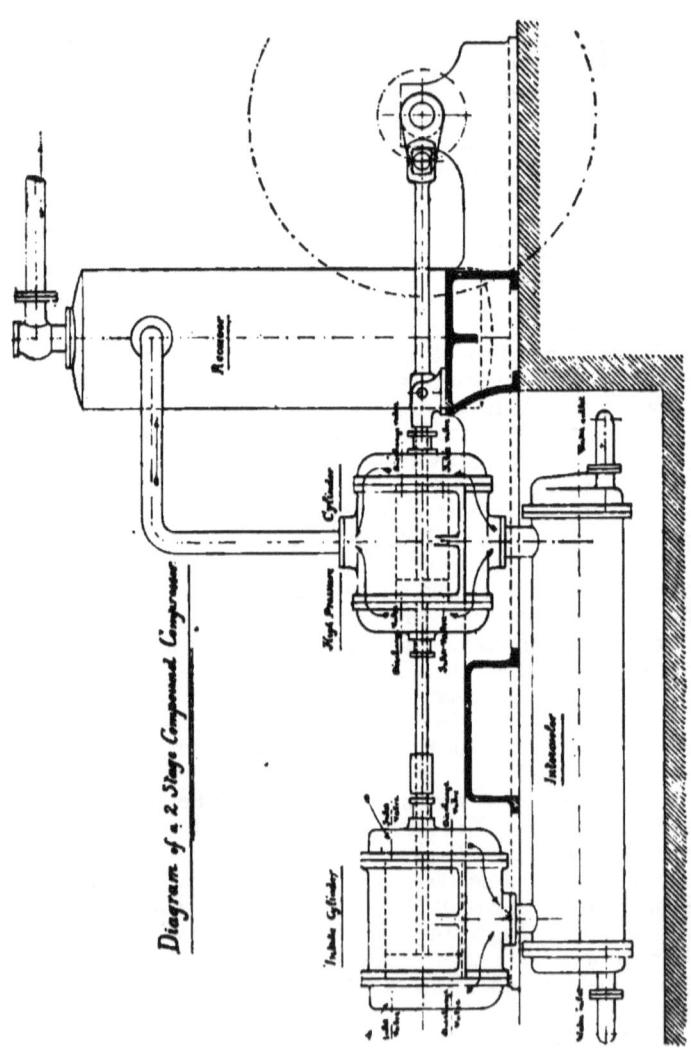

FIG. 55.—Diagram illustrating Compound Compression.

flywheel will regulate the motion of the machine than is the case in a single-stage compressor. For instance, if we use a 12-inch cylinder to compress air to 100 lbs. gauge, in the single-stage compressor, the load on the piston during one stroke will vary from 0 to 11,300 lbs., whereas in the compound machine this load can be made to vary from 0 to 5960 in all.

The principle of the Compound Compressor applies to any number of successive stages, and, theoretically, the more stages there are used the nearer will the compression approach the isothermal. But, at a practical standpoint, the increased number of cylinders is, of course, objectionable, inasmuch as it makes a heavier and more intricate machine, which will cost more and necessitate more expenditure for maintenance. The frictional resistances also become greater with the number of cylinders, and it is, therefore, readily seen that there are some practical limitations in the use of this system.

It may be stated that for pressures not exceeding 200 and even 300 lbs. per square inch, there should not be more than two stages in the compression. Four stages is the limit which has not been thus far exceeded, even with air pressure reaching to 2000 lbs. per square inch, and even for these high pressures three-stage compressors are deemed amply sufficient.

On the other hand, the compound system would be an unnecessary improvement with low pressures. For 50 or 60 lbs. receiver pressure it is quite likely that the percentage of extra resistances would balance if not overcome the percentage of gain in cooling.

In general, the advantages of a compound system consist in that less heat is developed at each stroke of the piston, while the air under compression is exposed to a larger cooling surface than in a single-stage machine.

The diagram, Fig. 56, represents the theoretical adiabatic cards of a 12x16 single stage compressor and of a tandem compound 12 and 7¾x16, both compressing to 70 lbs. gauge. It also shows the expansion curve in a 12x16 steam cylinder developing with steam at 80 lbs. gauge the same work as the single stage compressor.

These cards do not show the variations of pressure of steam and air, but the variations of effective load on the piston rod of the three cylinders, and they will serve for a comparison of two direct-acting steam compressors—one in the single stage and one in the compound system.

We know already that the aggregate piston load in the compound is less than in the single machine and as the initial loads are 0 in both cases, the range of variation is less in the compound. This allows a reduction in the size of the piston rods. It will be noticed that the compound curve has a sharper rise, since the maximum load H. G. is reached at the point I of the stroke, while in the single cylinder this same load is only reached at the point J. The result of it is that during this portion of the stroke, which precedes the point of equal loads in the two compressors, i. e., the point of intersection of

132   THE RIX COMPOUND COMPRESSOR.

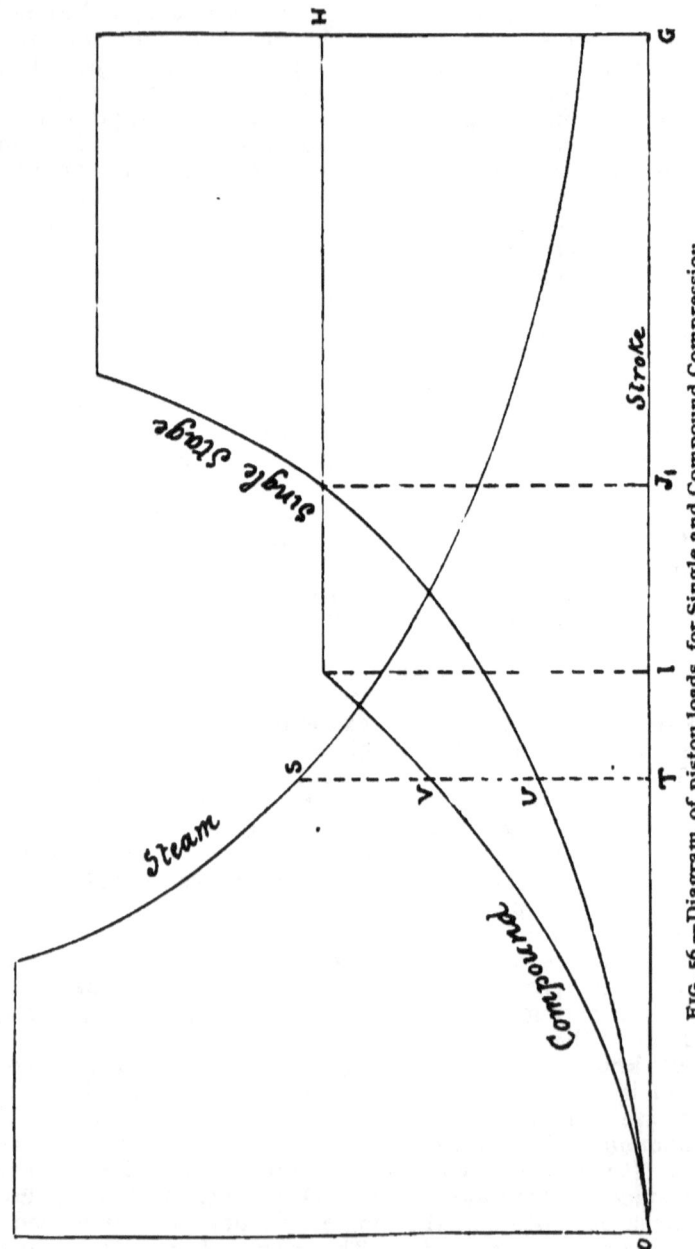

FIG. 56.—Diagram of piston loads, for Single and Compound Compression.

the steam and air curves, the difference of the load between the steam and air pistons is smaller in the compound, where it is $S\ V'$ for instance, than in the single cylinder compressor, where at the same point T, of the stroke, the difference is $S\ V$.

The same may be said for the second portion of the stroke, except in the region $I\ I'$, but here the discrepancy is unimportant, the piston loads being but little at variance in the two compressors, and this region corresponding to the maximum velocities of the pistons.

As the mass of moving pieces, whose momentum is resorted to for securing a regular motion, is a function of the actual difference between the steam air piston loads, lighter regulating pieces, like flywheels, will be required in the compound than in the single compressor.

The same size of steam cylinder will be found adopted in practise with both kind of compressors, the point of cut off being, moreover, variable.

A longer expansion of steam, combined with a less weight of machine, combine to win for a compound compressor the deserved claim of being a better balanced and more economical machine than the single stage.

It will be seen that a proper design of such machines must tend to an equal division of the total work among the several cylinders; that the loads are equal on each one of the pistons at any point of the stroke, and that the temperature of the entrance and exit of the air are the same in all the cylinders.

The following table shows the percentage of gain obtained by compounding as against the single-stage system, with various modes of compression:

PERCENTAGE OF GAIN OF 2-STAGE *vs.* 1-STAGE SYSTEMS OF COMPRESSION.

Ratio of Receiver pressure to atmospheric pressure..................	5	6	7	8	9
Gain per cent in:					
Adiabatic Compression (no cooling)...	11.5	12.8	13.8	14.8	15.9
Jacketed Cylinders....................	8.95	10.2	11	11.8	12.5
Jacketed Cylinders cooled by spray injection in the most efficient way possible	6.4	7.5	8.2	8.7	9.2

These figures show that for the usual air pressures the amount of work saved by compounding varies from 9 to 12 per cent. This is by no means a quantity to be neglected.

We also note that the advantage of compounding increases with the pressure and is more marked with a poor than with an improved system of cooling.

The Fulton Engineering and Shipbuilding Works do not issue a list of the various sizes of their Compound Compressors, for the reason that the relation between the two cylinders can never be fixed, the sizes of the initial cylinders depending of course upon the quantity of air required, and the size of the compound cylinders depending entirely upon the pressure desired. Special estimates and specifications are furnished with each compound machine. The following illustrations show some of the compound machines built by the Fulton Engineering and Shipbuilding Works, and give an idea of their general style.

The Compound Compressor, Fig. 60, shown in the preceding cut, illustrates the general style of the Compound Compressors built by the Fulton Engineering and Shipbuilding Works. This Compressor was built for the North Star Mining Company, of Grass Valley, Cal., and consists of Duplex Tandem Compound machines. The initial cylinders are 18 inches in diameter, and the high pressure cylinders are 10 inches in diameter by 24-inch stroke. The piston speed of the machine is 440 feet, which, while not quite as economical as one much lower, was dictated by the conditions under which the water wheel operated.

The air enters the initial cylinder at the temperature of the power room, which is approximately 62 degrees, and is therein compressed to 25 lbs. to the square inch gauge pressure. It leaves the cylinder at a temperature of 200 degrees Fahr. and passes through an intercooler of about 1000 running feet of 1-inch copper tubes placed directly beneath the water wheel, and which receives from the wheel a continual shower of water at a temperature of about 58 degrees. This cools the air to such an extent that it is delivered to the high pressure cylinders at a temperature of about 60 degrees. In these cylinders the air is compressed to 90 lbs. and is delivered from the cylinders at a temperature of 204 degrees into 6-inch mains, which lead to the mine. Indicator cards taken from the cylinders show that the cylinders are doing equal work, and at 110 revolutions they work smoothly and perfectly.

Notwithstanding the fact that some builders claim that clearance has no detrimental effect upon the economy of their air compressors, in the Rix compressors the clearance is practically eliminated, being not to exceed one-thirty-second of an inch at each end of the stroke. The cards taken from these cylinders are practically square-cornered.

The water-jacket system is quite unique, it being a duplex system—that is, there is an independent circulation for each end of the cylinder, the water passing longitudinally back and forth on the side of the cylinder and from the center in two

THE RIX COMPOUND COMPRESSOR. 135

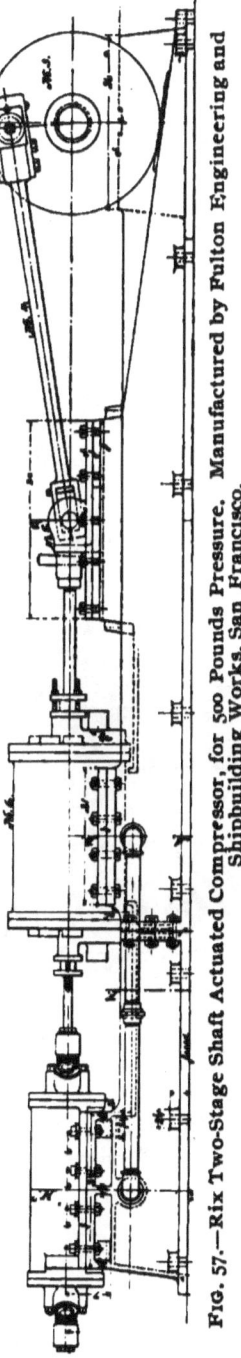

FIG. 57.—Rix Two-Stage Shaft Actuated Compressor, for 500 Pounds Pressure, Manufactured by Fulton Engineering and Shipbuilding Works, San Francisco.

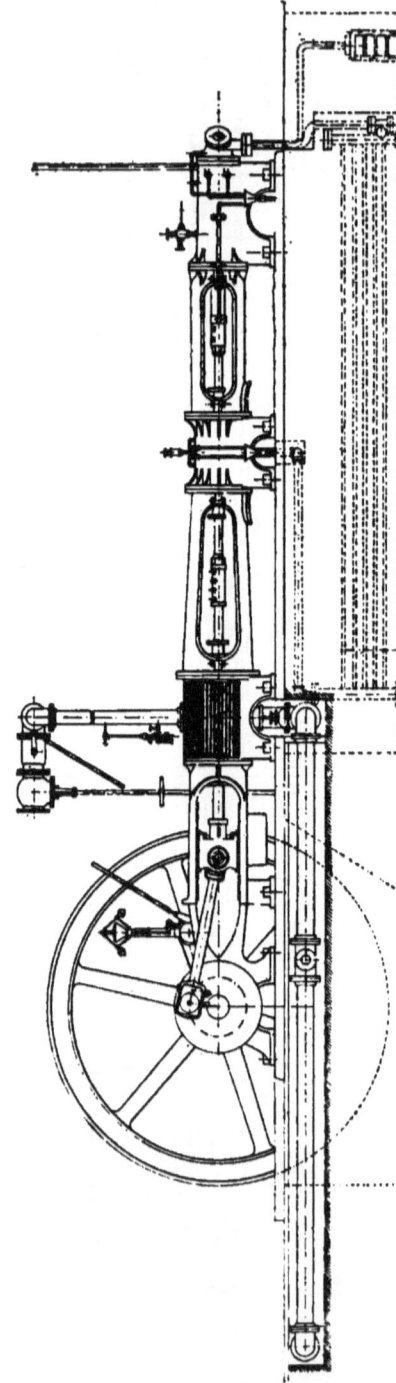

FIG. 58.—Rix Three-Stage Compressor, for 2000 Pounds Pressure. For description see article on "Pneumatic Torpedo Plant." Manufactured by Fulton Engineering and Shipbuilding Works, San Francisco.

THE RIX COMPOUND COMPRESSOR.

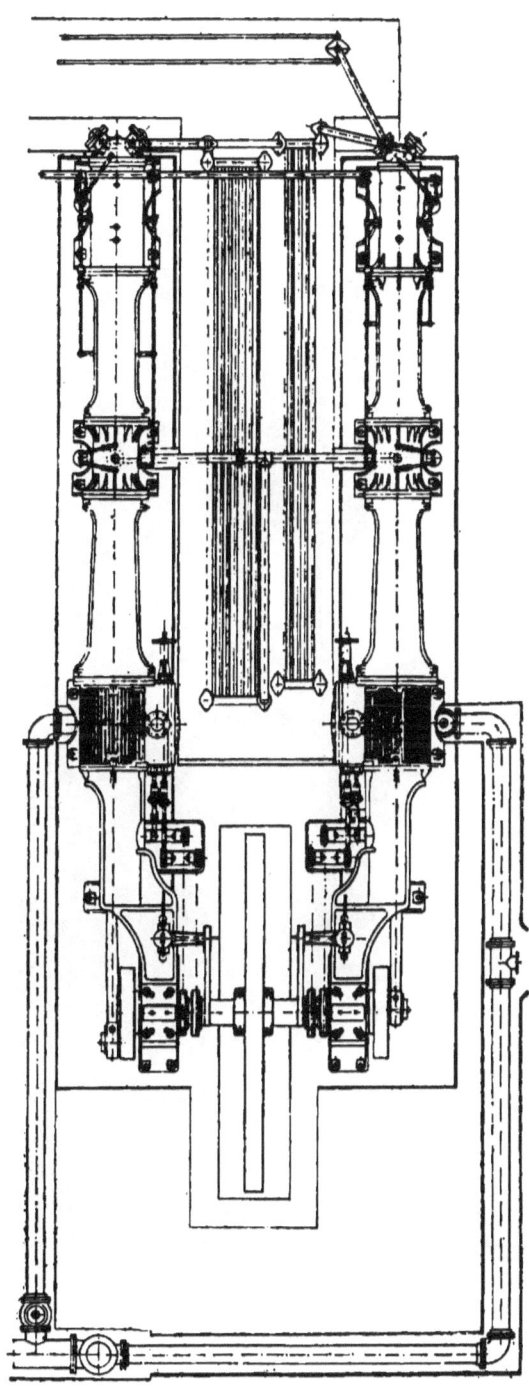

FIG. 59.—Rix Three-Stage Compressors, for 2000 Pounds Pressure. For description see article on "Pneumatic Torpedo Plant." Manufactured by Fulton Engineering and Shipbuilding Works, San Francisco.

138 THE RIX COMPOUND COMPRESSOR.

FIG. 60.—Rix Two-Stage Shaft Actuated Compressor. North Star Mining Company, Grass Valley, Cal. Manufactured by Fulton Engineering and Shipbuilding Works, San Francisco.

independent streams, cooling the heads at the same time. The efficacy of this water jacket will be noted in the temperatures above given.

In testing for volumetric efficiency, the receivers were carefully measured a number of times and found to contain 291 cubic feet. These were filled repeatedly, and the number of revolutions of the machine accurately counted each time. All of these experiments were conducted after the machine had been in operation for a sufficient length of time to reach its maximum temperature.

The barometer at the power house is 27.35 inches, corresponding to an elevation of about 2400 feet. This gives an atmospheric pressure of 13.32 lbs. per square inch. At 90 lbs. gauge pressure the ratio of compression would be 7.7, and the receiver containing 291 cubic feet represents 2240 cubic feet capacity of free air. The average of a great many experiments showed that the compressor took $102\frac{1}{2}$ revolutions to fill the receiver from 25 lbs, which is the pressure of the initial cylinder, to 90 lbs. At this pressure of 25 lbs. gauge there is 830 cubic feet of free air in the receiver. The difference between these two capacities, or 1410 cubic feet, would represent the amount of air which was forced into the receiver at the revolutions stated. Inasmuch as the temperature of the receiver is somewhat higher than the temperature of the inlet air, there should be a deduction made from this sum corresponding to that temperature of about two per cent, making the corrected amount delivered to the receiver 1382 cubic feet.

The theoretical capacity of the compressor, deducting the piston rods, and at $102\frac{1}{2}$ revolutions, is 1429 cubic feet of free air per minute. The ratio between 1382 cubic feet, actually delivered, and 1429 cubic feet, theoretical capacity, is 96.6 per cent, which represents the actual volumetric efficiency of the machine at the present writing. This of course will vary proportionately with the ratios of the absolute temperatures of the inlet air, depending upon the season of the year.

One peculiarity about the Rix Compressor, as may be noted from the cut, is the fact that the compressor is so arranged that any cylinder may be disconnected or any end of any cylinder may be disconnected without interfering with the operation of the machine. This feature is very valuable in case of repairs or accident to the machine.

To drive this compressor there has been placed upon the main shaft a Pelton water wheel, eighteen feet in diameter, which is believed to be the largest tangential water wheel ever made.

THE PNEUMATIC TORPEDO PLANT AT THE PRESIDIO.

(Originally published in "Journal of Electricity," S. F.)

The recent tests made by the military authorities on the dynamite guns at Fort Point may lend some interest to a few particulars regarding the Air Compressing Plant which forms the vital element of this installation.

The contract for the construction of the mechanical part of it, with the exception of the guns and their immediate fixtures, was awarded by the Pneumatic Torpedo and Construction Company of New York to the Fulton Engineering and Shipbuilding Works of this city, upon the plans and special designs of Mr. E. A. Rix, who supervised the construction of the plant.

The compression of air is made in three stages, from the atmosphere to the working pressure of 2000 lbs. effective per square inch. It is performed in two sets of horizontal engines, to both of which the subsequent description applies, they being in all respects entirely alike. The steam is supplied by four boilers of the Horizontal Tubular type, of 750 H. P. capacity, arranged to work either with natural or with forced draught.

Two steam cylinders connected to the same shaft by cranks at an angle of 145 degrees from each other, actuate in tandem, that is, through their piston tail rods, each two air cylinders, there being on one side one low pressure and the intermediate or second stage cylinder, and on the other side one low pressure and the high pressure or finishing cylinder.

This duplex set therefore comprises two steam cylinders, two intake cylinders, wherein the atmospheric air is compressed to about 75 lbs. effective, one intermediate cylinder, carrying the air pressure from 75 to about 400 lbs. effective, and one high pressure cylinder, which takes the air at 400 lbs. and compresses it to 2000 lbs. effective.

The intake or low pressure cylinders are double acting, that is, they have inlet and discharge valves at each end, while the intermediate and high pressure cylinders are single acting, that is, provided with valves at one end only, their pistons being plunger rams with spherical heads, connected to the tail rods of the intake cylinders.

The special purpose which these compressors have to serve made their design and construction subservient to conditions at entire variance with the lines upon which an air compressing plant is usually established. The main object of the designer, when a large power is to be used, as in the case of the Fort Point installation, is commonly to secure the greatest possible economy in the production of the compressed air. In the present instance, compound condensing engines of the most approved type, and air cylinders working at a moderate linear

Pneumatic Dynamite Gun at Fort Winfield Scot, San Francisco.

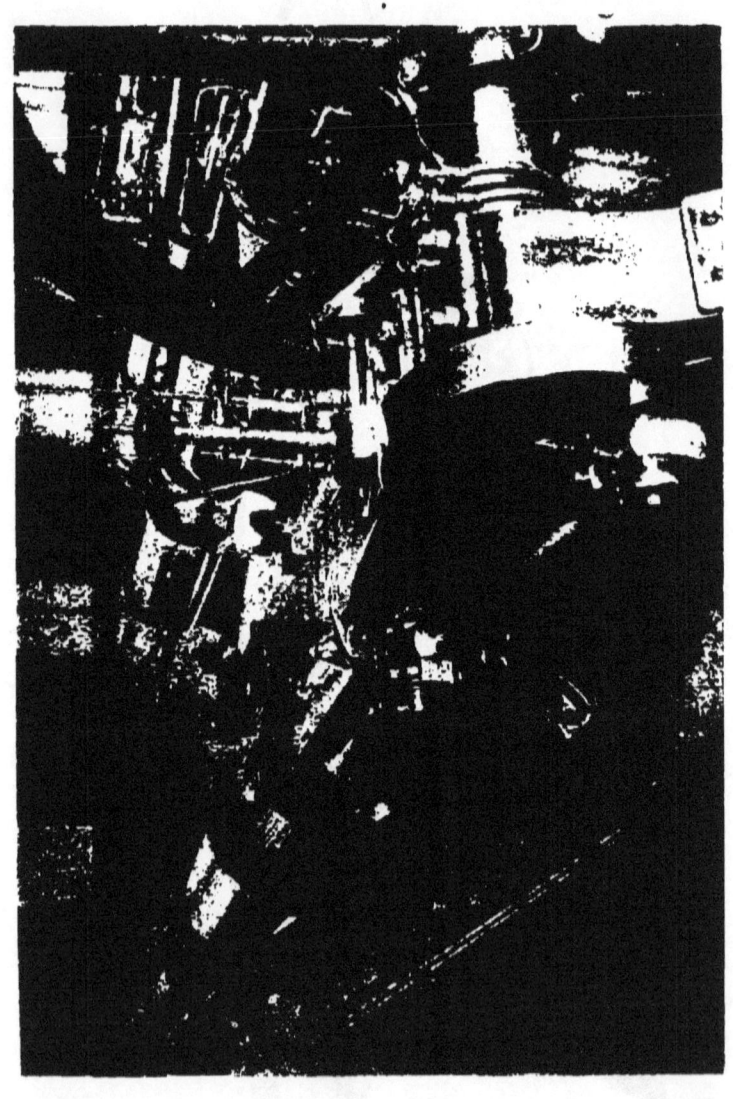

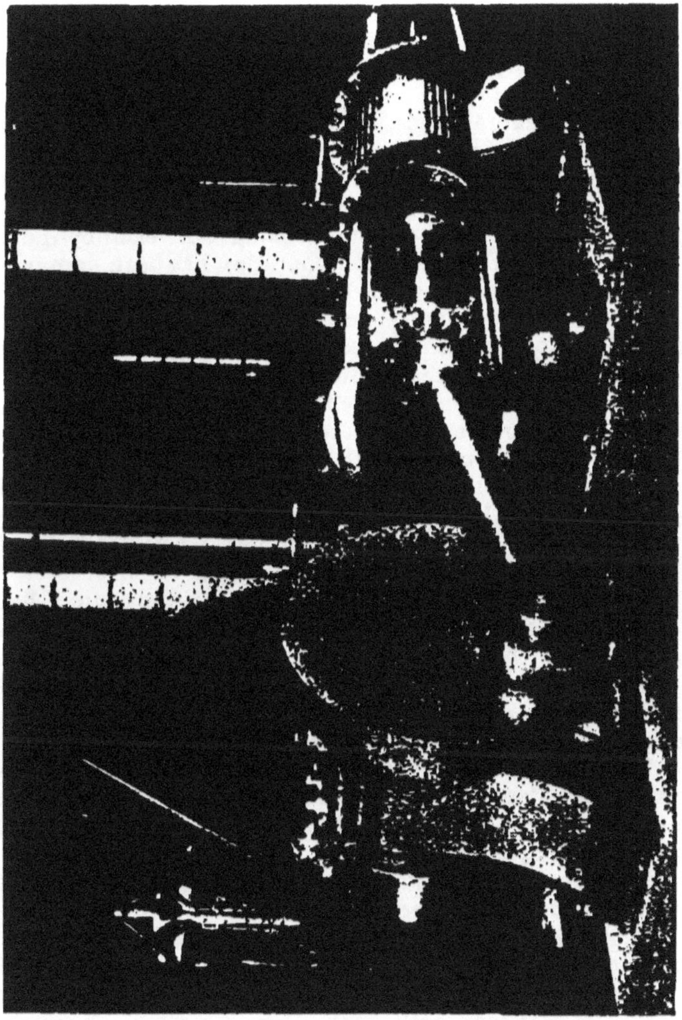

THE PNEUMATIC TORPEDO PLANT. 143

Meyer's Cut-off Engine, Rix Air Compressors. Dynamite Gun Plant.

piston speed, would present themselves to the mind as advisable. Such engines would be established in view of a regular working speed, or approximately so, and everything would be provided to give the economical appliances a chance to work to their full advantage.

At Fort Point the primary requirement was to have a plant as little liable as possible to getting out of order. Solidity, simplicity, and endurance were therefore the main points to be considered, economy being a desirable but decidedly an accessory feature.

Upon these general lines, supplemented by conditions of capacity within a given time, of efficiency in the means of cooling the air and of practical effectiveness of several important parts, the present plant was designed, built, and erected.

The steam engines are non-condensing and each cylinder acts independently; that is, no compounding has been adopted. The valves are provided with Meyer's cut-off, regulated by hand, the Governors merely acting on the throttle in case of racing. The cranks are set at the angle heretofore indicated, in order that the machine may be balanced as nearly as possible and yet the engines be able to start in any position.

In the air cylinders the greatest care has been used to secure a cooling efficiency as high as possible. The heads and the barrels of the cylinders are water-jacketed, the water discharge pipes from the jackets being in full view and easily accessible, and the supply of cooling water being regulated according to its temperature at the discharge.

A very elaborate and effective system of intercoolers has been established between the intake and intermediate cylinders and also between the intermediate and high pressure cylinders. These intercoolers consist of nests of copper pipes extending under the floor in cemented trenches, where a stream of cold water is constantly running. The proportions of these intercoolers have purposely been made very ample, and their effectiveness is fully demonstrated by the low temperature of the air before it enters the intermediate and the high pressure cylinder, which are given hereafter.

A similar cooler is provided for the air at working pressure after it leaves the high pressure cylinder and before reaching the 24 forged steel storage tubes, which through a complete system of pipes and manifolds, and also a compact arrangement of valves, can be set in communication with each particular gun, or if so desired, with a supplementary storage supply located in the foundation of the guns.

That the demand upon the compressors may vary during action, within widely distinct limits, was exemplified by the fact that while 360 feet per minute is generally considered as a limit of piston velocity in water jacketed cylinders, this velocity

THE PNEUMATIC TORPEDO PLANT. 145

Initial Cylinder, Rix Air Compressors. Dynamite Gun Plant.

Intermediate Pneumatic Ram, Rix Air Compressors. Dynamite Gun Plant.

has been, during part of the trials, carried to 568 feet, or an excess of 58 per cent. At this high rate of speed no undue heating could be observed in the moving parts and the absence of jarring and of trepidations was the best evidence of the remarkable strength and steadiness of the plant.

Of course, when working at high speed, no claim is nor could be entertained to maintaining a satisfactory cooling efficiency in each individual cylinder. As before stated, the intercoolers are of sufficient size to deal with the heat liberated during the compression even at high speed. But when the period of compression, and, of course, the period of effective possible cooling, lasts two-fifteenths of a second, the heat units passing through the cylinder walls during that time cannot be expected to be many. It might be argued that the Riedler compressors in Paris work at a nominal piston velocity of 550 feet and occasionally 733 feet per minute, but aside from the fact that the use of a spray for cooling and of mechanically moved valves are both combined to reduce the rise of temperature, the pressures in the two-stage Riedler compressor are considerably lower, the air being sent into the mains at only 118 lbs. gauge per square inch, an insignificant pressure as compared to 2000 lbs.

Another point of interest in the Fort Point plant is the absence of leakage at the stuffing boxes of the intermediate and high pressure rams. This point has been the cause of much annoyance in similar plants built elsewhere, and the present arrangement is the outcome of long and costly experiments.

The friction, in a running joint capable of holding 2000 lbs. of air pressure against the atmospheric, is necessarily enormous, and after the nature, the shape, and the size of the packing had been determined upon, it became necessary to keep the packing sufficiently cool to prevent its rapid wear. This is effected by a special circulation of cold water inside the rams, the arrangement being quite apparent on the general plan, and that it is successfully effected can be easily ascertained. This water circulation also partly contributes to cooling the air under compression.

At the normal rate of speed of about 400 feet per minute of piston velocity, the compressors supply to the storage tubes 460 cubic feet of air per hour at 2000 lbs. gauge. The annexed abstract from trials made in view of timing the production of the compressors gives interesting evidence of the effectiveness of the intercoolers and of the regularity of the temperature of air at its entrance to each cylinder.

For a range of final pressures comprised between 800 and 2000 lbs. effective, the variation of temperature was only 8 degrees Fahr. for the intermediate and 3 degrees Fahr. for the high pressure cylinder, the temperature of the engine-room being 71 degrees Fahr.

148 THE PNEUMATIC TORPEDO PLANT.

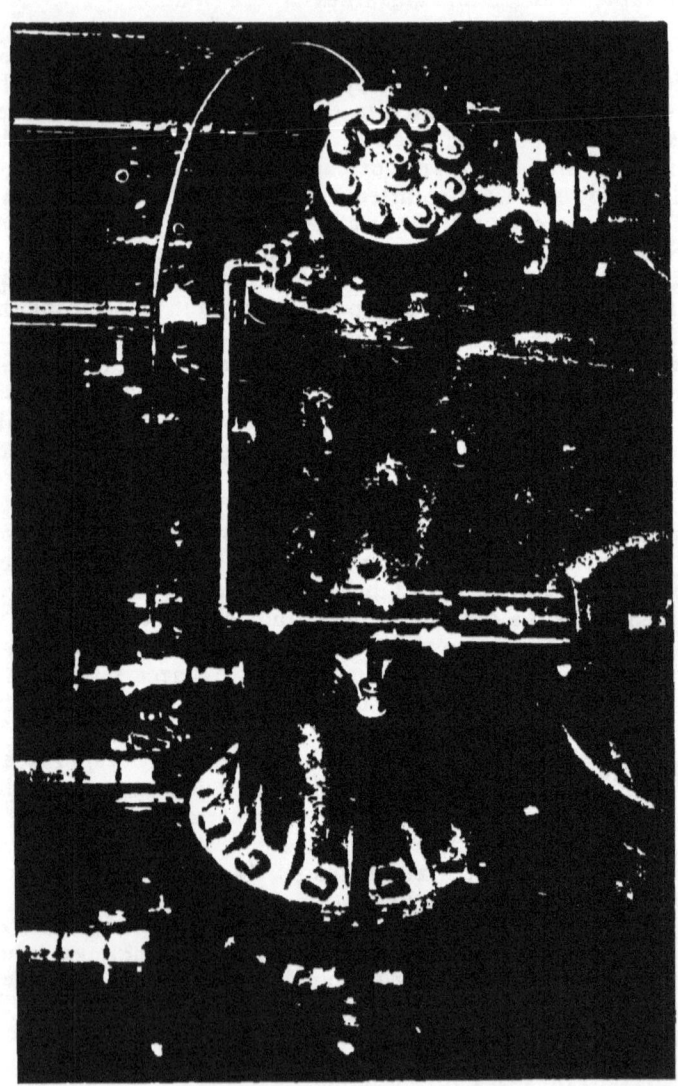

High Pressure Pneumatic Ram, Rix Air Compressors. Dynamite Gun Plant.

THE PNEUMATIC TORPEDO PLANT.   149

High Pressure and Intermediate Rams, Rix Air Compressors.   Dynamite Gun Plant.

Gauge pressure lbs. per sq. in.	Fahr. temperature at entrance to		
	L. P. Cylinders.	I. P. Cylinders.	H. P. Cylinders.
800	71	67	66
900	71	68	67
1000	71	69	67
1100	71	69	67
1200	71	70	68
1300	71	70	68
1400	71	71	68
1500	71	72	68
1600	71	72	68
1700	71	74	69
1800	71	74	69
1900	71	73	69
2000	71	72	69

The discharge temperature of the low pressure cylinders gradually increased and then remained stationary at 320 degrees Fahr. The intermediate cylinder discharge likewise attained a temperature of 292 degrees Fahr., and the high pressure cylinder, beginning at 375 lbs. per square inch, and at a temperature of 66 degrees Fahr., delivered from the intercoolers, gradually rose in temperature as the pressure increased, until it reached 2000 lbs., and after running at that pressure for one hour, the thermometer indicated its maximum, viz., 358 degrees Fahr.

The sum total of those temperatures, viz., 970 degrees, as compared to the adiabatic temperature of single stage compression to 2000 lbs., which is 1762 degrees Fahr., indicate the work saved by the three-stage method of compression combined with the jacket and ram cooling devices.

The compression throughout the whole range was practically regular, being as an average 115.1 lbs. for each 500 revolutions of both machines.

The mean of many cards taken from the steam cylinders showed that each compressor absorbed 342.61 I. H. P., while the cards from the three air cylinders showed 293.78 I. H. P. for each compressor. The work then absorbed by the friction, inertia, etc., was 48.83 I. H. P. or 14.2 per cent of the indicated power employed, showing a mechanical efficiency for the compressor of 85.8 per cent, which is high, especially in view of the facts that the engines were new and consequently stiff to some extent, and also that some extra friction is developed at the ram stuffing-boxes as compared with a compressor working at the usual air pressures.

The resisting load of 48.83 H. P. while the compressors were doing full duty may be compared with the friction load on the machine without air pressure, and an interesting result

THE PNEUMATIC TORPEDO PLANT. 151

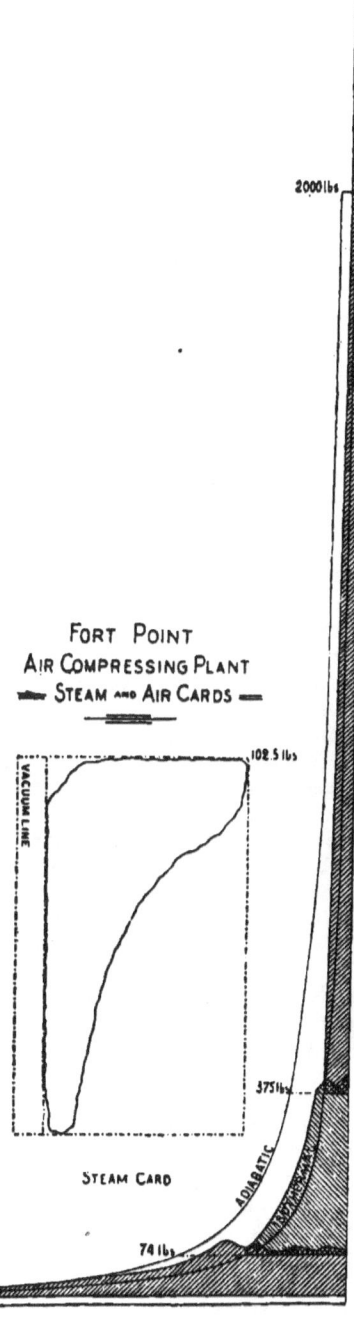

obtained. Cards taken showed that this friction load was 32.4 H. P., being .663 of the resisting work under load and showing an increase of 50.7 per cent in the resistances between no load and full load.

The combined indicator cards illustrated herewith are plotted from actual cards and show a saving of 36.8 over adiabatic single stage compression.

The boilers for this plant were of the Return Tubular type, and manufactured by the Chandler & Taylor Co. of Indianapolis, Ind.; were 72 inches in diameter, by 16 feet long, and of a nominal horse power of 500, which were increased by the forced draught employed, to about 750 horse.

These boilers were tested to 150 lbs. to the square inch, and fully satisfied the requirements of the Treasury Department. The forced draught was employed because it was not considered desirable to continue the stacks above the roof, and thus give an opportunity for invading forces to discover the position of the plant. A short stack was therefore necessary, about fifteen feet in length, which required the employment of a forced draught. The forced draught was instituted by two Sturtevant fans, with engines attached, having cylinders three inches in diameter by three and a half inch stroke. These fans delivered each 12,000 cubic feet per minute of free air, through a 22-inch main, which, passing underneath the battery of four boilers, was connected to each by a 10-inch outlet underneath the grate bars. It was found during the test that these fans need be run only to about 60 per cent of their capacity.

The engines exhausted their steam into two heaters of the National type, of 300 H. P. each, which furnished to the boilers feed water at a temperature of 200 degrees Fahr.

The Feed Pumps were of the Deane type, being Duplex and two in number, the steam cylinders being six inches, the water cylinders being four inches, and the stroke being six inches. At a slow piston speed these pumps furnished all the necessary water, which was drawn from the pits after being heated by the air from the compressors.

As an auxiliary there are installed alongside of the Feed Pumps two Nathan Injectors of 300 H. P. each, which are amply sufficient to furnish all of the water necessary to feed the boilers.

During the test for rapidity of firing, while the plant was supposed to be strained to its utmost, the firemen had ample time to observe the operation of the compressor plant, showing that the boilers were more than sufficient to supply the steam necessary for the proper operation of the compressors.

The electrical plant was furnished by the Electrical Engineering Company of this city, and consisted of one 35-kilo-Watt compound wound dynamo, capable of being worked up to 25 per cent of its rated capacity for thirty minutes without undue heat, and operated by an Armington & Sims engine.

This dynamo was connected by about 800 feet of two-wire, insulated copper cable, encased in lead covering, and capable

THE PNEUMATIC TORPEDO PLANT. 153

Armington & Sims Engine. Driving Dynamo. Dynamite Gun Plant.

of carrying a current of 400 amperes, without undue heating. This cable was placed in and fastened to the side of an underground conduit.

This Company also placed in position at about ten feet distant from the dynamo, a switchboard of slate, and wired complete, having three double-pole three hundred ampere knife switches.

The compressed air, after leaving the compressors and being confined in the storage tanks, was distributed to the three guns independently, through a manifold of bronze, having attached five gauges, two registering 2000 lbs., and three 1250 lbs., and so arranged with valves that any or all of the guns could be operated at once.

This air is carried to the underground storage reservoirs of the guns, through a pipe having an outside diameter of $2\frac{1}{2}$ inches, and inside diameter of $1\frac{5}{8}$ inches and duly tested to 3500 lbs. to the square inch for tightness.

From the guns to these manifolds also there are three copper pipes, $\frac{1}{4}$ inch inside diameter by $\frac{1}{2}$ inch outside diameter, to register the pressures at the manifolds that are contained in the carriages of the guns.

This is in general the description of the air-compressing plant. We now come to speak of the guns themselves, which were manufactured at the West Point foundry on the Hudson, each 15 inches in diameter, with a length of 50 feet; each gun mounted on its carriage, weighing about 70 tons, perfectly balances, and these are mounted upon concrete foundations.

The tests of these guns for their mechanical efficiency, which may be called their ease of operation, showed that they could be traversed by the electric motors, which were situated in the gun carriage, in an average of one minute, throughout the entire 360 degrees, and they could be elevated from extreme elevation to extreme depression, in from eight to eleven seconds. Any one familiar with the length of time necessary to operate ordinary powder guns by hand will appreciate the fact that this facility of operation is marvelous.

For testing these guns for mechanical efficiency, the requirements were, first, that 45 shots should be fired in the first hour and 30 shots in the hour succeeding. Inasmuch as the wastage of air would be the same whether actual projectiles were fired, or whether the air was simply wasted through the muzzle of the gun in "air shots," no projectiles were fired in this test, and it was found for the first hour that 45 shots were fired and the compressors running at their normal speed registered a final pressure of 1800 lbs., it being thus demonstrated that the compressors were amply sufficient to maintain any requirements which might be placed upon the gun. Twenty air shots were fired to ascertain the utmost rapidity with which

THE PNEUMATIC TORPEDO PLANT. 155

35 K. W. Dynamo for ranging Dynamite Guns.

they could be discharged, and the same were discharged in 7½ minutes, though the contract did not require that these shots should be discharged inside of 30 minutes, it being thus demonstrated that the compressors and the guns were amply capable to maintain the test required by the Government.

The test for rapidity of firing with actual projectiles took place next. The projectiles used were pieces of gas pipe 12 inches in diameter and 8 feet long, loaded with sand. The weight was 1040 lbs. Each one of the three guns was required to fire five of these projectiles within twenty minutes. The test developed the fact that these projectiles were all discharged from each gun within eight and one-half minutes, and they were by far the most interesting feature of the whole test.

Having no means for maintaining the accuracy of their flight, these projectiles were nevertheless thrown for the first one-half distance of their flight perfectly accurate; that is, they maintained the position of a well-directed projectile, after which they tumbled end over end and fell into the sea. Without any plain table measurements being taken upon them, they apparently fell quite accurately within a small target.

The time of flight of these projectiles averaged about nineteen seconds for about 2200 yards.

The question of rapidity of firing and of loading having been determined, the next test was one of accuracy, and the live projectiles were discharged from these guns at a distance of 5000 yards. The projectiles used were of the eight-inch caliber, the difference in diameter being made up by wooden pistons in four sections so that the wooden pieces would fly off after the projectile had left the gun, leaving it free to make its flight. The first projectile flew 5000 yards and exploded; the second projectile flew 5070 yards and exploded; the third projectile flew 5015 yards and exploded; the fourth projectile flew 5040 yards and exploded; all of these projectiles being plotted on a plane table in a rectangle 70 yards long by 20 yards wide, the time of flight being about 27½ seconds.

As a matter of experiment, two shots were fired into the hills of Marin County, at a distance of 3350 yards, each with the 8-inch sub-caliber shell loaded with 100 lbs. of dynamite, the first shot being fired five days previous to the second shot. The shots struck within 45 yards of each other and exploded in a perfectly satisfactory manner; in fact, the pits caused by the explosion joined each other. The larger shells, viz., the 15-inch full caliber projectiles being eleven feet long and weighing some 1050 lbs., loaded with 500 lbs. of nitro-gelatine, were thrown into the sea at a range of an average of 2100 yards. They exploded practically upon striking the water, throwing into the air a column of water about 100 feet in diameter at the base, and, from the levels taken at the gun, about 400 feet in altitude.

The tests as above enumerated were perfectly satisfactory in every respect and exceeded in every way the requirements of the Government. There were no mistakes made and no delays whatever caused by the air-compressing plant or the gun plant, which probably exceeded the Government requirements in an aggregate of over one thousand per cent, if the various exceed percentages of the different tests were added together, and which reflected great credit upon the manufacturers of the power plant, the constructing engineer, the manufacturers of the guns and projectiles, and also the Pneumatic Torpedo & Construction Company of New York, which contracted for and thus successfully carried to completion their contract with the Government.

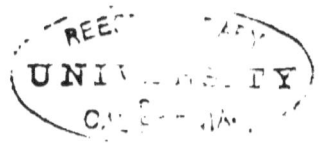

ROCK DRILLS.

The RIX and the GIANT ROCK DRILLS are manufactured in San Francisco, Cal., and their construction is the result of a study of the requirements of the Pacific Coast in rock drilling, covering the last twenty years. It has been the aim of the manufacturers of these machines to produce something which will be especially satisfactory to the miners of the Pacific Coast.

Many of the improvements in these machines have been suggested by the operators of the drills themselves, to suit particular conditions, and it has been the aim of the manufacturers to construct a machine which is rapid and powerful in its action.

The GIANT and the RIX DRILLS are manufactured under the following patents, controlled by EDWARD A. RIX.

U. S. PATENTS AS FOLLOWS.

Re-issue **6,705**   Patent No. **190,699**
Patent No. **149,013** Patent No. **206,067**
Patent No. **152,712** Patent No. **235,296**
Patent No. **156,003** Patent No. **235,816**
Patent No. **169,389** Patent No. **255,335**
Patent No. **172,529** Patent No. **410,334**
Patent No. **178,214** Patent No. **454,228**
         Patent No. **490,152**

Others pending.

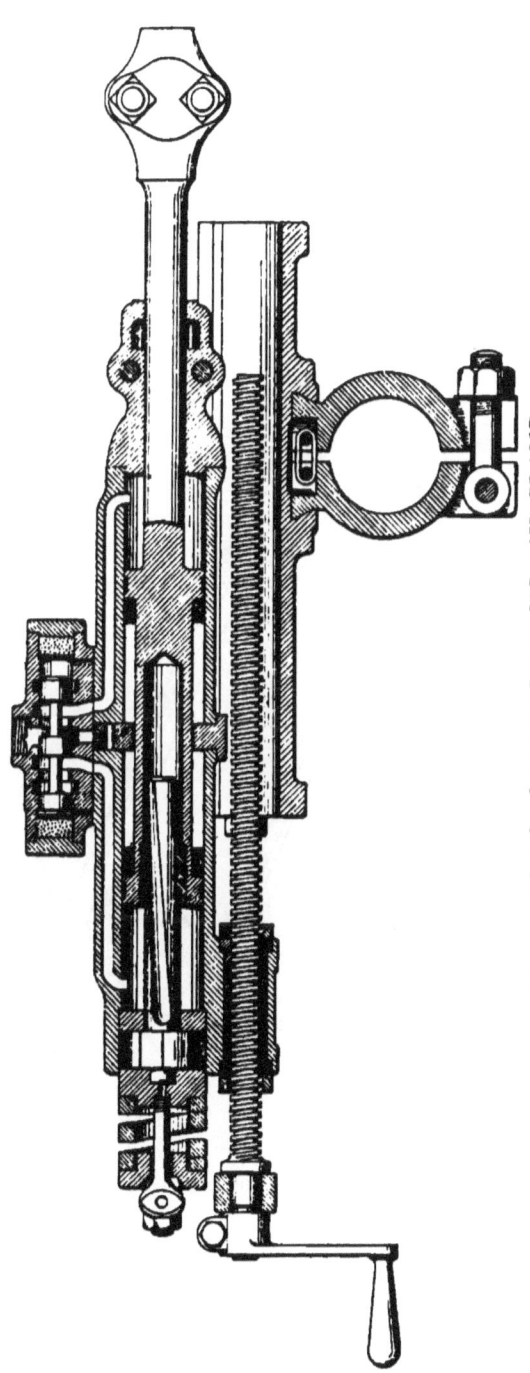

Fig. 61.—Sketch of RIX ROCK DRILL AND CLAMP.

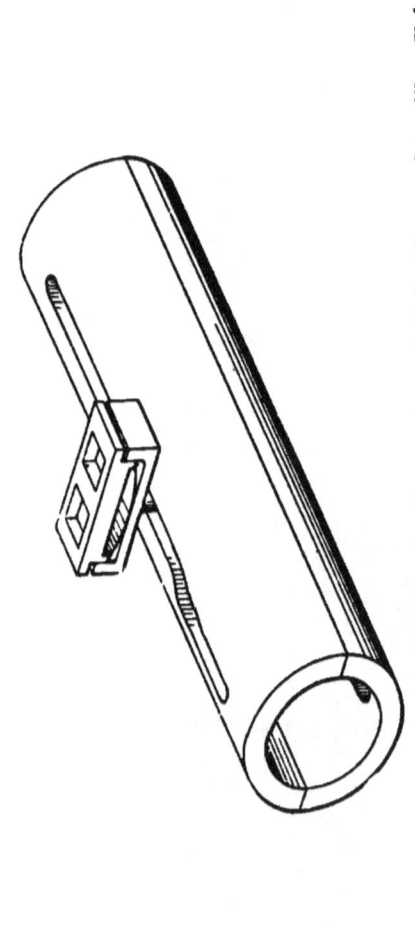

Fig. 62.—RIX DRILL VALVE MOTION, showing the Cam, the Valve Claw, and Auxiliary Valve.

Knowing that the average man who runs a rock drill is not a skilled mechanic, in the construction of the RIX DRILL the aim has been to produce a valve motion which could not by any means whatever go wrong or fail to go together, providing no piece should be omitted. To accomplish this the entire VALVE MOTION is arranged symmetrical to a line perpendicular to the line of motion and passing through the center of the exhaust. This permits the main valve spool, the caps, plates, and buffers, the auxiliary valve, the whole valve chest, or anything pertaining to it, to be reversed in any way, and the result is a proper and complete valve motion, and it also allows the exhaust to be turned in any direction by simply using an ordinary street elbow.

All JOINTS on either of these drills are scraped, and there are no gaskets to get out of order.

One of the most annoying faults about imported rock drills is the rubber buffer, which has to be introduced into both heads in order to prevent accident to the heads by the careless operators. Especially is this true when steam is used, for the rubber rapidly disintegrates and interferes with the proper working of the machine. In both the RIX and the GIANT DRILLS these interior buffers are dispensed with and a SPIRAL SPRING is placed on the back head of the machine which does service for both heads and which never wears out. In fact, a duplicate spring has never been furnished for any of these machines. A flat bow-spring does not accomplish the same result, as it breaks quite readily, and is generally replaced by a solid bar to avoid further difficulty.

Quite a feature with the GIANT and RIX DRILLS is in the use of the same sized COLUMN, CLAMP, and TRIPOD for any of the machines. The result is that a mine need purchase but one sized mounting, and any drill will fit thereon. A 3-inch drill may be taken out of the heading if hard rock is encountered and a larger machine attached to the same clamp at once, without any re-setting of the column, and this is also found especially valuable in upraising work.

All of the machines above the 2¾-inch size use the same hose and the same COUPLINGS, and any of the machines will take drill steel of any size up to 1¼-inch, and use any shape bushing.

The above-named conveniences are of great consideration, and have never failed to commend themselves to intelligent purchasers. It may be urged that a COLUMN which is large enough in diameter to properly carry a 3-inch machine is too small for a 3½-inch drill. This may be true where the machines stand away from the column to any extent and where they are being racked by lost motion and where they reciprocate slowly, but with the GIANT and RIX machines, which hug the column closely and which have no lost motion on account of the DOUBLE FEED NUT DEVICE, and which

162   ROCK DRILLS.

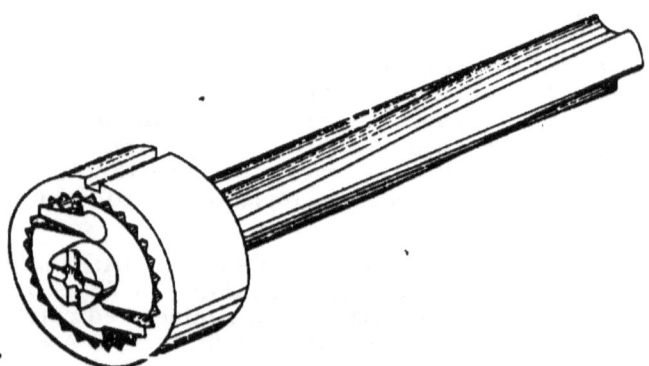

Fig. 63.—RIX ROTATING DEVICE. Patented.

reciprocate fully fifteen per cent faster than any other drill, it is not necessary, and therefore a purchaser need not pay for that which he does not require.

The ROTATING MOTION in these drills is one of the finest features about them, and it fulfils perfectly every requirement. From the sketch, it will be seen that it consists of an internal ratchet engaging with swinging pawls carried in the head of the rotating bar. A very slight spring pressure serves to throw them into contact, when by the nature of the angles of adjustment the pawls will be carried into a pinch that cannot slip or be broken. All the angles in the ratchet and pawls are right angles; therefore the ratchet may be reversed after it is worn on one side, and equal service be given to the other.

The same is true of the PAWLS, and being symmetrical, it does not matter which side or end is first presented for duty.

This feature of having nearly all of the moving parts symmetrical and reversible is quite a feature in the construction of these machines and is of immense assistance in the cost of operating and convenience, as well as being very useful in emergency.
It is not necessary that these PAWLS SHOULD BE REVERSIBLE,—a fact which has been taken advantage of by an Eastern drill manufacturer—and the owners of the patent on this rotating device desire us to state for them and in their behalf that the INTRODUCTION OF THIS SWINGING PAWL IN A DRILL ROTATING MOTION, WHERE THE PAWL IS SYMMETRICAL OR NON-SYMMETRICAL, IS AN INFRINGEMENT UPON THEIR RIGHTS, AND ANY PARTIES USING SAME WITHOUT PROPER LICENSE FROM THESE ORIGINAL PATENTEES WILL BE ENJOINED FROM USING SAME AND BE ALSO REQUIRED TO PAY DAMAGES.

All rock drills, of either the RIX or the GIANT pattern, which use compressed air as a motive power, are supplied with a FRONT HEAD, which has no stuffing box but which is internally packed with a leather-cupped ring, which is absolutely perfect in its action. This is an old method of packing a drill piston rod, having been used about twenty years ago, and is now used by other drill makers occasionally. It has, however, never given any great amount of satisfaction and never was absolutely tight, for the air had always escaped through the split in the ring, and the cup was not the proper shape.

The LEATHER CUPS, however, for these drills are made by a machine especially constructed to shape the joints, forming a perfect interior and exterior cylinder, one-eighth of an inch apart. There is no split at all, and they remain perfectly tight under any pressure and last about four months under continuous wear.

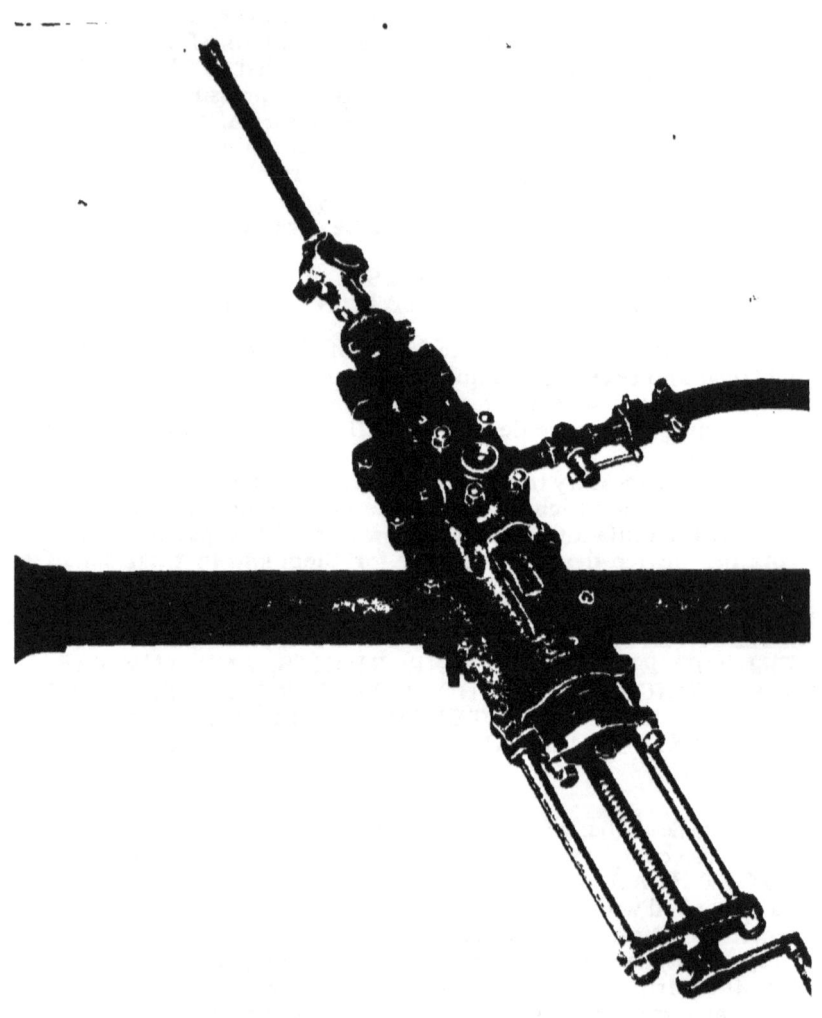

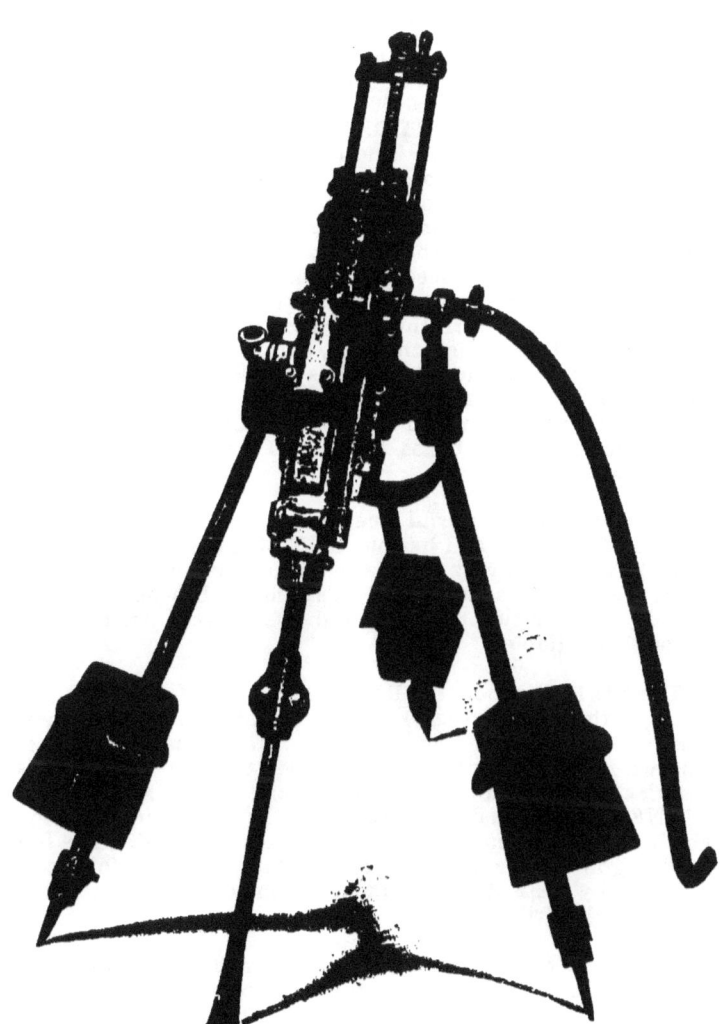

FIG. 65.—GIANT DRILL, Mounted on Tripod.

The FEED NUT DEVICE is another special feature of both of these machines. All other rock drills are provided with a single feed nut, and this together with the feed screw naturally wears rapidly. After wearing so that the lost motion becomes apparent, it acts materially against the cutting power of the machine, as well as being noisy and a fruitful source of accidents, for at every stroke of the ordinary rock drill it thumps back and forth in its cage to the full extent of this lost motion. The only remedy is a new nut and screw.

In the RIX and GIANT DRILLS, by means of the double feed nut all trouble of this kind is avoided. One of the nuts is secured to the cylinder of the drill in a manner similar to all drills; the other has a toothed edge and may be turned to the extent of a tooth at a time as the feed screw wears. This allows the front edge of the feed screw thread to work on the back edge of the first feed nut thread, and the back edge of the feed screw thread to work on the front edge of the second feed nut thread, thus furnishing the feed screw with practically one perfect-fitting nut all the time, and in this manner, a feed screw may be worn until its threads break away without any lost motion being apparent in the drill. It needs no comment to show that the drill uses FEW FEED SCREWS, in fact, the life of the screw is not less than TWO YEARS in any case, barring accident.

The clamp is a powerful one, very light, a perfect DROP STEEL FORGING, and has but ONE BOLT, so that it is easy to work, and being very light can be operated in half the time that it requires for some others. This clamp has been in continuous use for twenty years and has proved itself to be thoroughly reliable.

The PISTON of both the RIX and the GIANT DRILLS is so arranged that it will receive any size bushing up to $1\frac{1}{4}$ inches. The drills are always fitted with an octagon bushing, unless otherwise ordered, for that style receives the steel just as it is manufactured and thus saves the expense of TURNING THE SHANKS as well as doing away with the annoying breakage which happens when the ends of the steel are turned. The full size octagon is none too strong to withstand the powerful blows delivered by these machines and a much lower pressure must be used if the drill shanks are turned.

The COLUMN MOUNTINGS used for these machines are similar to those used by other makers, excepting that only one size is manufactured. Other sizes are made and kept in stock to satisfy the ideas of customers who have been used to other drills, but the increased size is not necessary to a satisfactory working of the machines.

The TRIPOD is one furnished with universal joints to its

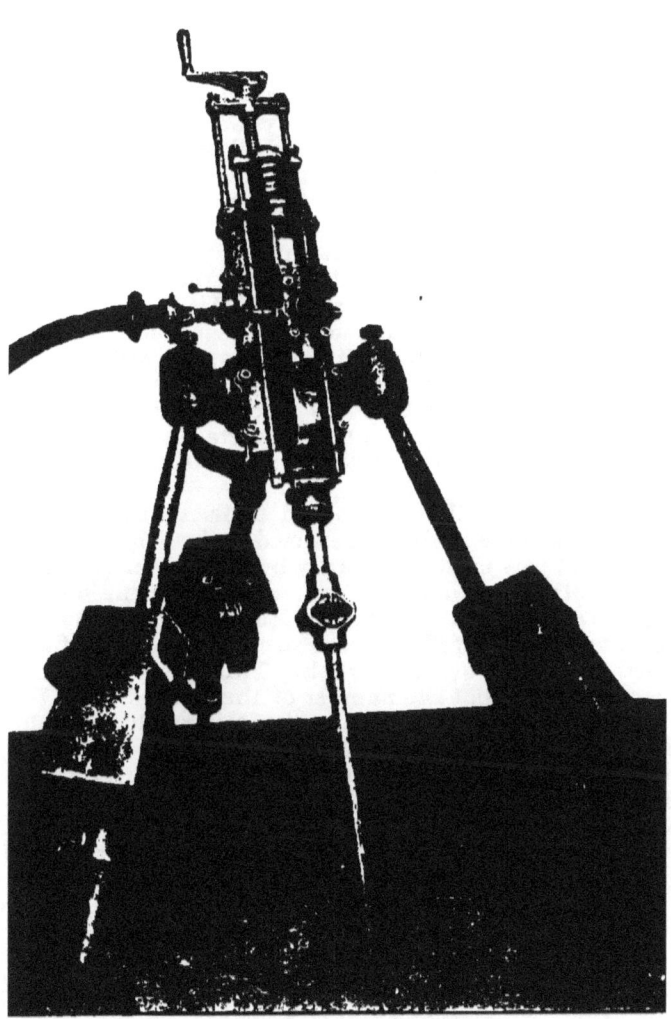

FIG. 66.—2¾-inch RIX DRILL, Mounted on Tripod.

legs and has stood the test of twenty years of good service. A clamp is always used with this tripod; this enables the head of the tripod to be used as a short column, so that the drill may be given a lateral motion of about four inches, a feature which is very useful in drilling holes in uneven rock full of cracks or fissures, or which from any cause deflects the drill steel.

In the RIX DRILL the VALVE MOTION is in every way superior to anything which now operates a rock drill, and one of the most noticeable things about the drill when it is running alongside of other makes, is the wonderful regularity of its motion reciprocating as evenly as a steam engine, and delivering a blow with much greater velocity than any other machine, and also more of them. Most of the drill makers make a claim for an uncushioned blow, and that the valve does not change until after the blow is struck, but these are not facts which are consistent with another claim which they make; viz., that their machines are the only ones which make a variable stroke. None of the standard makers claim that the reversing of the valve is dependent upon the striking of the rock, yet their statements would lead one to that conclusion. Every one knows that the drills will run at quite a speed without striking even the front head, and any one who examines their valve mechanism will perceive that it is practically the same for both the front and back stroke, and they certainly would not like the inference drawn that the piston must strike the back head in order to reverse the valve.

The fact is that all the standard drills strike a cushioned blow, and the valve is always reversed before the drill strikes the rock, and this must necessarily be so in order to allow for a variable stroke, and to provide for a sufficient number of strokes. Drills have been used in Europe, and many experimental ones made here have been so constructed that the valve changed after the blow was struck. This, undoubtedly, gives the heaviest blow, but the number of the strokes is so limited that can be delivered in a minute, that the machine could not begin to do the work an ordinary rock drill can do. The more the cushion in a drill, the faster it will reciprocate, and the less effective will be the blow. The less the cushion, the heavier the blow and the less the number of strokes. The shorter the working stroke, the greater the number of strokes and the less the blow, and the less the working pressure, the less the number of strokes and the less the force of the blow. Therefore, in fashioning a rock drill, the result must be a mean between these four relations, which shall give the best results. In other words, the length of stroke, the amount of cushion, the number of strokes, and the pressure used, must be so adjusted with relation to each other that the best result will be produced—allowing, of course, that the diameter of the cylinder has been determined. All these problems have been very satisfactorily solved in both the RIX and the GIANT machines.

The VALVE MOTION of the GIANT DRILL is one which is operated directly from the piston by mechanical contact, and this drill is manufactured to satisfy the beliefs of some drill users that a machine of this construction is better than a machine operating with the auxiliary valve motion, such as the RIX.

The sizes of the GIANT DRILL are made to alternate with the sizes of the RIX, so that the following Tables of sizes and capacities, which represent a complete range, are offered to the public:

Descriptive Table of Rix Rock Drills

Letter indicating Size	A	B	C
Diameter of Cylinder Inches	2 3/4	3 1/4	3 5/8
Length of Stroke Inches	6 1/4	7 1/4	8 1/4
Extreme Length of Drill over all	3' 1"	3' 6"	3' 8"
Diameter of Supply inlet Inches	3/4	1	1
Weight of Machine Lbs	175	285	345
Weight of Tripod complete Lbs	580	580	580
Strokes per Minute. 60 Lbs effective Pressure	500	500	500
Length of Feed Inches	24	25	27
Depth of Vertical Hole Machine will drive easily Feet	8	15	20
Diameter of holes that Machine will drill Inches	1 1/4 to 2	1 1/4 to 2 3/4	1 1/2 to 3
Diameter of Steel used Inches	1 to 1 1/8	1 1/8 to 1 1/4	1 1/4 to 1 3/8
Size of Boiler required H.P.	8	10	12
Size of Supply Pipe up to 200 feet Inches	1	1 1/4	1 1/4

Descriptive Table of Giant Rock Drills.

Letter indicating Size		D	E	F
Diameter of Cylinder	Inches	$2\frac{3}{4}$	$3\frac{1}{8}$	$3\frac{1}{2}$
Length of Stroke	Inches	$6\frac{1}{4}$	$6\frac{1}{2}$	$7\frac{1}{2}$
Diameter of Supply Inlet	Inches	$\frac{3}{4}$	—	—
Weight of Machine	Lbs	170	275	335
Strokes per Minute, 60 Lbs Effective Pressure		350	350	350
Length of Feed	Inches	24	25	27
Depth of Vertical hole Machine will drive easily	Feet	8	15	20
Diameter of holes that Machine will drill	Inches	2	$2\frac{3}{4}$	3
Diameter of Steel used	Inches	1	$1\frac{1}{8}$	$1\frac{1}{4}$
Size of Boiler required	H.P.	8	10	15
Size of Supply pipe up to 200 feet	Inches	1	$1\frac{1}{4}$	$1\frac{1}{4}$

DUPLICATE PARTS OF THE RIX ROCK DRILLS.

1—Rotating Nut.
2—Piston, bare.
3—Piston Ring.
4-5—Sleeve.
6—Feed Nut (adjustable).
7—Feed Nut (plain).
8—Yoke for Feed Nuts.
9—Lower Head.
10—Leather Crimp for Lower Head.
11—Chuck Bolts and Nuts.
12—Chuck Bushing.
13—Chuck Key.
14—Steam Chest, bare.
15—Main Valve.
16—Steam Chest Cap.
17—Steel Cushion Plate.
18—Rubber Cushion.
19—Auxiliary Valve.
20—Auxiliary Valve Spring.
21—Auxiliary Valve Claw.
22—Oil Screw.
23—Yoke for Head Bolts.
24—Head Spring.
25—Cover for Ratchet Ring.
26—Bottom Plate for Ratchet Ring.
27—Rotating Bar.
28—Cylinder, bare.
29—Guide Block.
30—Shell Strip.
31—Cylinder Bolts.
32—Shell Bolt.
33—Feed Screw.
34—Yoke for Shell Bolts.
35—Feed Screw Handle (brass).
36—Pawl.
37—Ratchet Ring.
38—Pawl Spring.
39—Shell without Strips or Yoke.
40—Clamp Wrench.
41—Steam Chest Wrench.
42—Chuck Wrench.

ROCK DRILLS. 173

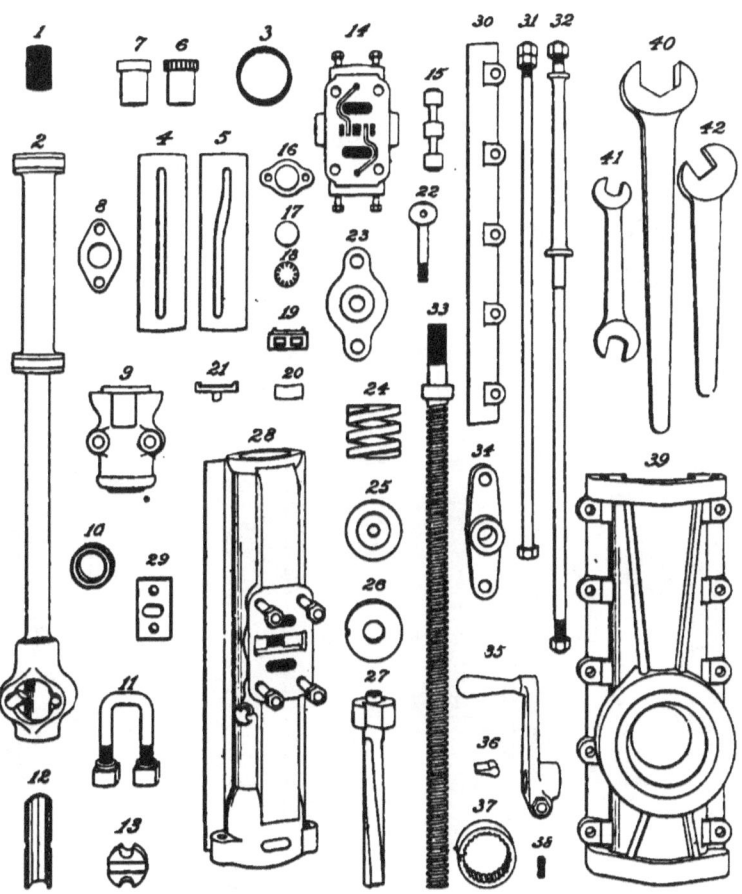

DUPLICATE PARTS OF THE RIX ROCK DRILL.

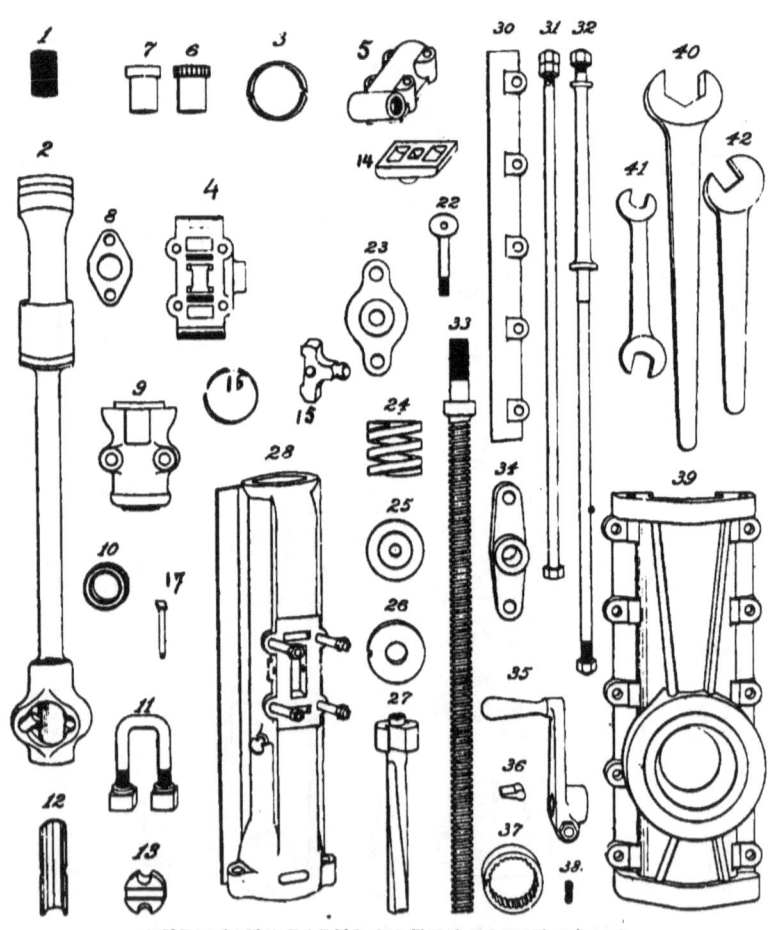

DUPLICATE PARTS OF THE GIANT DRILL.

DUPLICATE PARTS OF THE GIANT ROCK DRILLS.

1—Rotating Nut.
2—Piston, bare.
3—Piston Ring.
4—Valve Chest.
5—Valve Chest Cover.
6—Feed Nut (adjustable).
7—Feed Nut (plain).
8—Yoke for Feed Nuts.
9—Lower Head.
10—Leather Crimp for Lower Head.
11—Chuck Bolts and Nuts.
12—Chuck Bushing.
13—Chuck Key.
14—Valve.
15—Valve Rocker.
16—Piston Ring Spring.
17—Rocker Pin.
22—Oil Screw.
23—Yoke for Head Bolts.
24—Head Spring.
25—Cover for Ratchet Ring.
26—Bottom Plate for Ratchet Ring.
27—Rotating Bar.
28—Cylinder, bare.
30—Shell Strip.
31—Cylinder Bolts.
32—Shell Bolt.
33—Feed Screw.
34—Yoke for Shell Bolts.
35—Feed Screw Handle (brass).
36—Pawl.
37—Ratchet Ring.
38—Pawl Ring.
39—Shell without Strips or Yoke.
40—Clamp Wrench.
41—Steam Chest Wrench.
42—Chuck Wrench.

RIX PLUG AND FEATHER DRILL.

The Rix Plug and Feather Drill, a cut of which appears in Fig. 66½, is the smallest drill manufactured by this Company. It has a two-inch diameter cylinder, from four to five inch stroke, and makes from seven hundred to nine hundred strokes per minute. It is designed for drilling small holes about one inch in diameter and for depths up to twenty-four inches.

For quarry work it is mounted on a tripod, as shown in the cut, and for mining purposes it has the usual column mountings. The tripod is one which gives a wide range of movement.

The Drill itself weighs about 65 lbs. and is extremely convenient to handle. It is generally used with seven-inch steel and the chuck is made tapering to take the end of the steel in similar to the way a twist drill fits in its socket. This will be found most convenient in the handling of these small drills.

This machine will be found very handy for many ranges of work, including the driving of wooden pins in caison, scow, or dry dock constructions where the pins have to be driven from underneath the work being constructed.

In the use of air it is very economical, taking about twenty-five cubic feet of free air per minute.

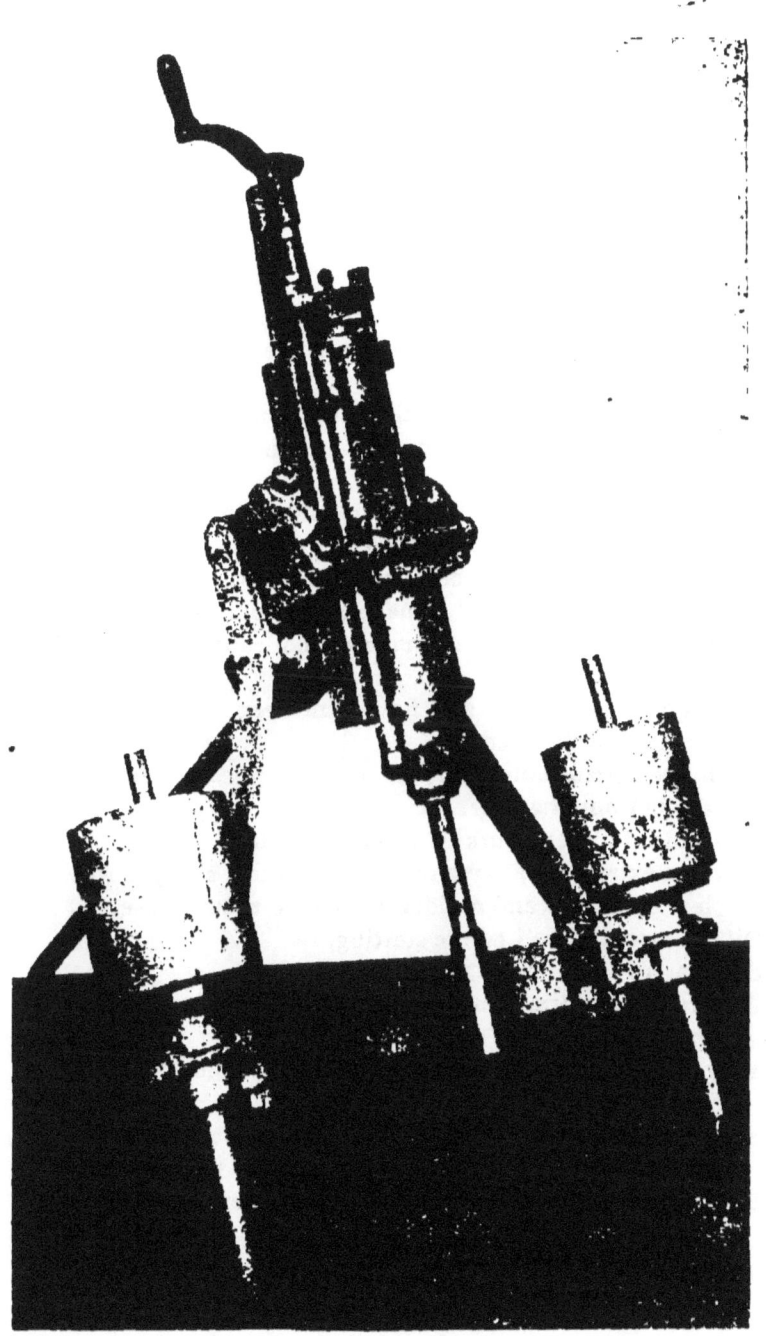

FIG. 65½.—RIX PLUG AND FEATHER DRILL.

A FEW GENERAL HINTS.

Buy a Compressor larger than you need.
Buy one which is economical.
Run it slow.
Put in good foundations.
Have a spare boiler if you can afford it.
Have a clean, ship-shape engine-room.
Cover all of your steam-pipes.
Provide large air-pipes.
A generous sized receiver will come in handy.
Make as few short turns as possible in the air-pipe.
Use a good cylinder lubricant.
Circulate ample water in the air cylinder jackets.
Have some extra compressor valves, and change them frequently.
Put in one or two shut-off valves in your air-pipe.
Keep the receiver properly drained.
Buy a rock drill of a size best suited to the work, and don't buy any unless your mind is made up to do it properly.
Have plenty of steel, so your men are not running for drill-bits all the time.
Get a good blacksmith, and have him keep both ends of the steel properly sized.
Drill good-sized holes, for the powder does better work at the bottom of a hole.
Have an intelligent workman to run the drill.
Have an extra drill always ready in the shop, and you will find less breakages and accidents occur to those in use.
Oil the machine well before starting.
See that all the nuts are tight.
Be sure that no dirt is in the hose before it is attached to the machine.
Keep the column well jacked up, and have blocks of wood top and bottom.
Start the holes on the shortest stroke of the machine, and gradually lengthen out the stroke as the hole deepens.
Feed the machine so that the piston will clear the front head.

In soft ground, make haste slowly.

If the steel gets stuck in the hole, strike it sharply until it releases.

Never strike the chuck.

Do not screw up too hard on the chuck-nuts or clamp-bolts, for it is perfectly possible to break them.

Keep your bushings in good order.

A bit of cast-iron or iron borings thrown into a fissured hole will help it out.

A piece of broken drill-bit will often cause a hole to run out.

Drill wet holes whenever you can.

A leaky stuffing-box will often prevent the piston pulling out from a tight hole.

Never run the drill against the head to throw the steel out.

Do not expect the drill to furnish brains to run itself.

Do not expect it to run without repairs.

Carry as high a pressure as possible when your rock is hard, and calculate always that the repairs will vary, as the pressure and also the work done.

Remember that a rock drill is an engine, after all, and the fewer times it goes over the dump, or is dropped off the column, or is blasted upon, the longer it will last.

Generous and faithful oiling will help a machine wonderfully.

Use a good steam-trap when using a drill in a quarry.

A tripod must be securely set to do good work.

The same kind of drill-points do not work equally well in different kinds of rock.

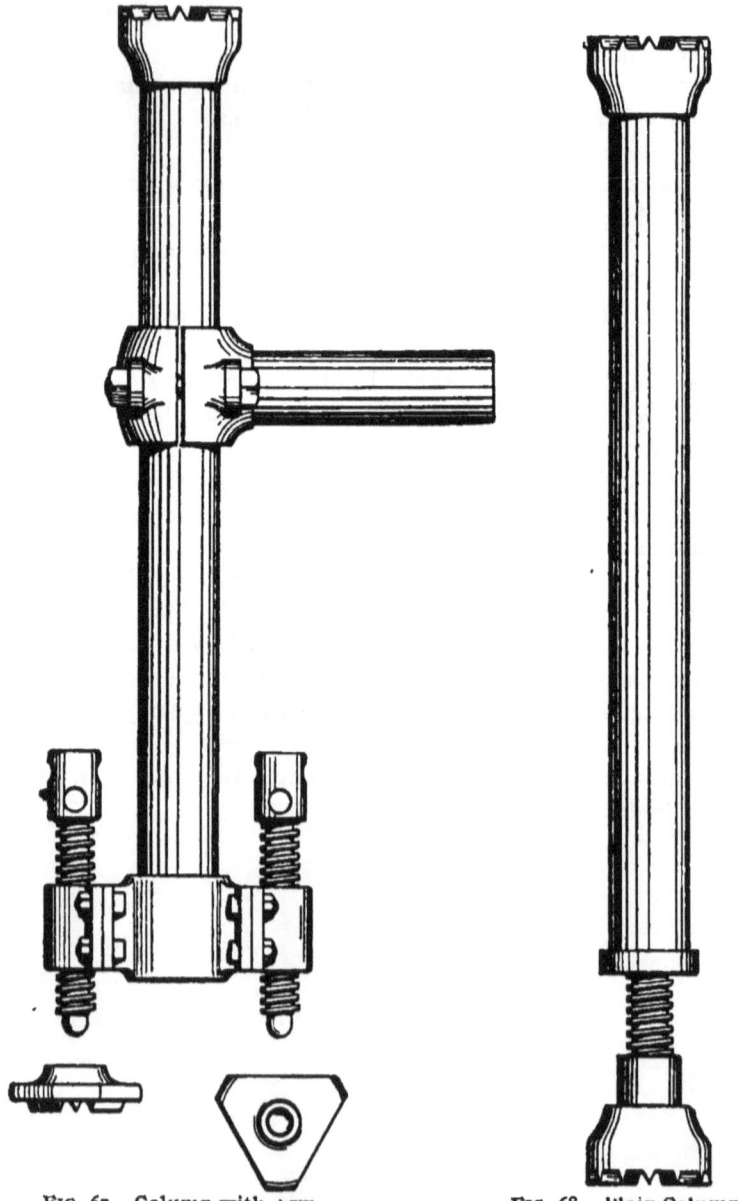

FIG. 67.—Column with Arm.   FIG. 68.—Plain Column.

Column Mountings for Rock Drills. Made in any length. One price for all lengths under ten feet.

ROCK DRILLS.

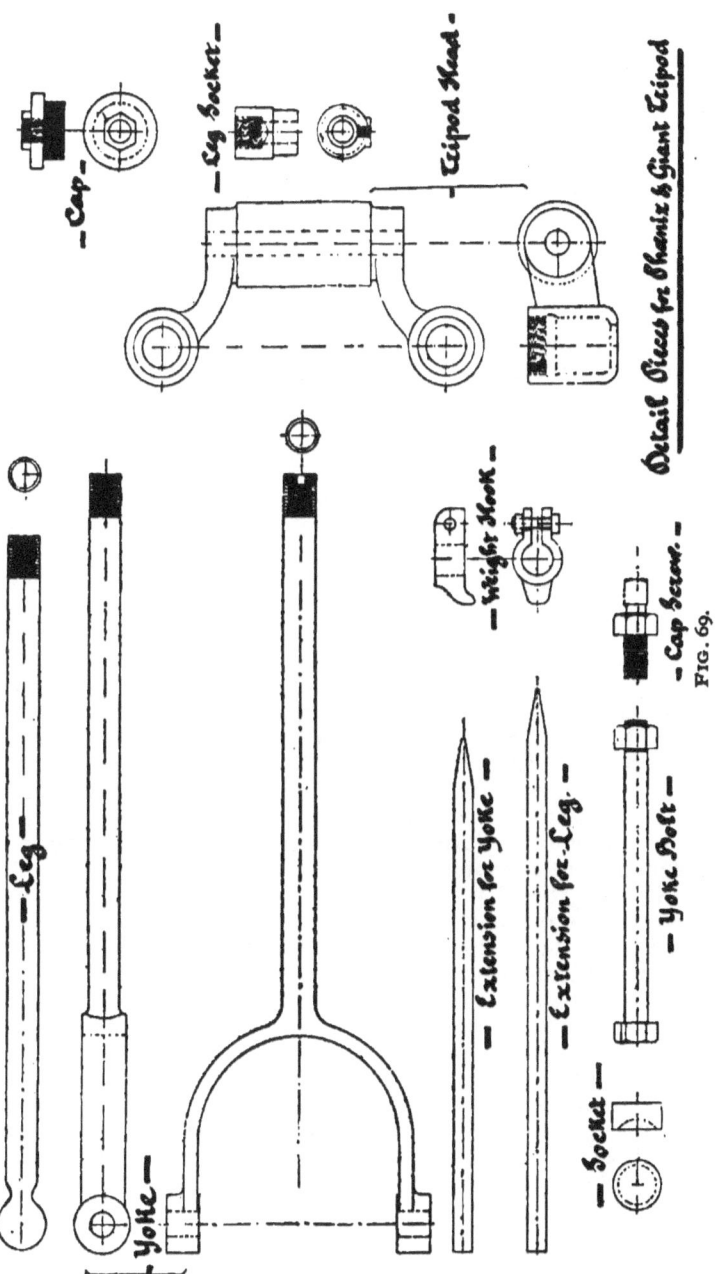

FIG. 69.

AIR RECEIVERS.

In conjunction with an air compressor there is generally attached a reservoir called an air receiver. The purpose of this is twofold: to collect the moisture which is condensed from the air after it is compressed, and also to afford a sufficient volume to receive the intermittent discharges from the compressor, and reduce them to a continuous flow in the pipes leading from the receiver.

The ordinary receiver is fitted with an air gauge, a safety valve, and a valve to draw off the moisture. These are arranged as shown in the cut herewith attached.

Our reservoirs are made of homogeneous steel, with bumped heads, of a sufficient thickness to be tight at 125 lbs. cold water pressure, for all ordinary plants. We prefer bumped heads because bracers are not then necessary. We put three cast iron feet on one end of the receiver for it to stand upon, and sufficiently high to permit drawing off the entrained water water easily, above the floor line.

We are frequently asked where is the proper place for the receiver—at the compressor or in the mine? We reply, *both*. There never was too much receiver capacity on any plant. We do not believe it essential to have a very large receiver near the compressor, providing there is an opportunity to place one further along the pipe. About fifteen times the cylinder capacity would, in all ordinary cases, keep the gauge steady at the compressor. It would be a great benefit to systems having medium or small size pipes to have as large a receiver capacity at or near the point where the air is used, and especially is this the case where hoisting engines are drawing from the air pipes. It requires no engineering knowledge to see that if air receivers could be made large enough to diffuse the intermittent work into an average draw on the pipe leading from the compressor, that the compressor need be only large enough for the average work, whereas ordinarily it must be large enough for the maximum work, and consequently uneconomical.

It is not generally practicable to have reservoirs so large, however, but a reasonable approach can be made to this capacity without much expense. We have known compressors to do 25 per cent more useful work by putting receivers near the point where the air is to be used, and where numerous bends and elbows are required in the main pipe.

When air is drawn too fast through the main pipe, causing a reduction of pressure, the increase of volume due to the loss pressure causes quite a marked increase in all the frictional losses through the system. We therefore advise receivers at both ends of the line, the smaller ones near the compressor, and this is independent of the amount of storage capacity in the pipe.

DIMENSIONS OF AIR RECEIVERS.

Diameter, inches	30	30	36	36	36	42
Height, feet	6	8	8	10	12	8
Thickness of Shell, inches	$\frac{1}{4}$	$\frac{1}{4}$	$\frac{1}{4}$	$\frac{1}{4}$	$\frac{1}{4}$	$\frac{1}{4}$
Thickness of Heads, inches	$\frac{5}{16}$	$\frac{5}{16}$	$\frac{3}{8}$	$\frac{3}{8}$	$\frac{3}{8}$	$\frac{3}{8}$
Weight	700	900	1200	1400	1600	1800
No. of 3¼-inch Drills Receiver is suitable for	1	1	2	3	4	5

Diameter, inches	42	42	42	48	48	48
Height, feet	10	12	16	10	12	16
Thickness of Shell, inches	$\frac{1}{4}$	$\frac{1}{4}$	$\frac{1}{4}$	$\frac{5}{16}$	$\frac{5}{16}$	$\frac{5}{16}$
Thickness of Heads, inches	$\frac{3}{8}$	$\frac{3}{8}$	$\frac{3}{8}$	$\frac{7}{16}$	$\frac{7}{16}$	$\frac{7}{16}$
Weight	1900	2000	2100	2400	2900	3400
No. of 3¼-inch Drills Receiver is Suitable for	8	10	12	12	15	20

184 AIR RECEIVERS.

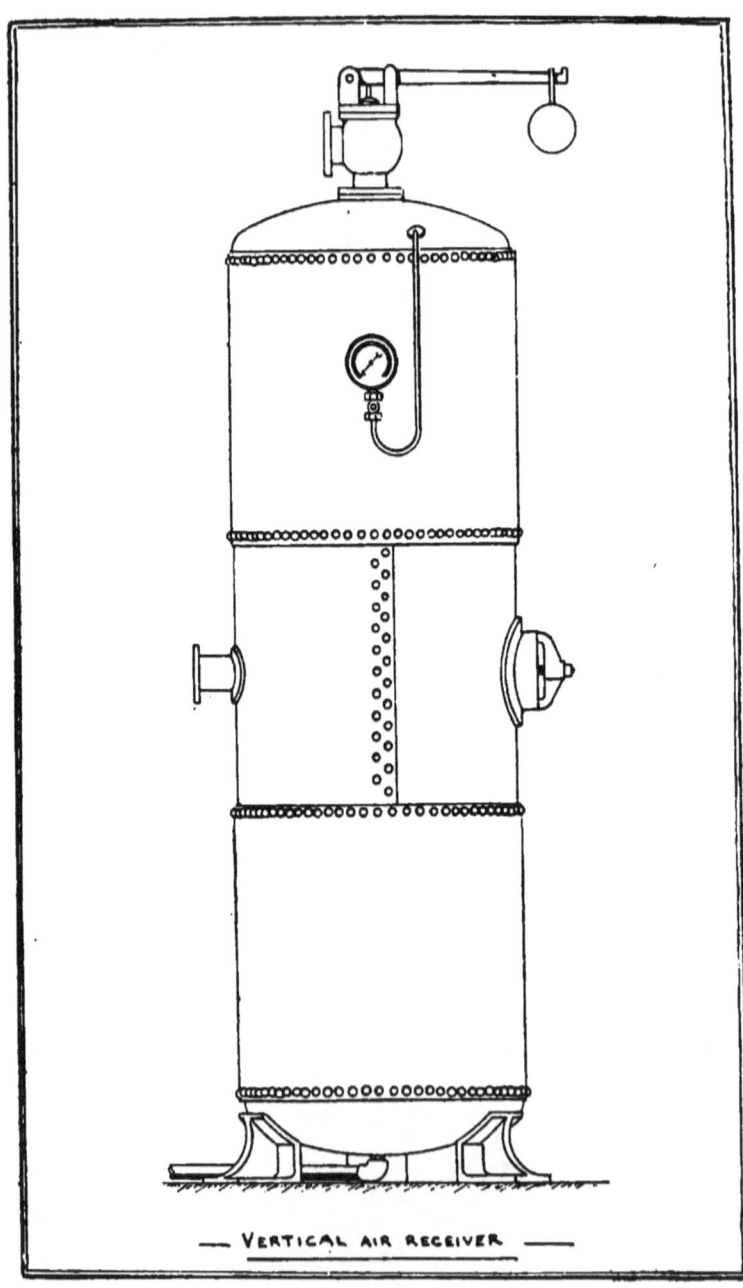

VERTICAL AIR RECEIVER

FIG. 70.

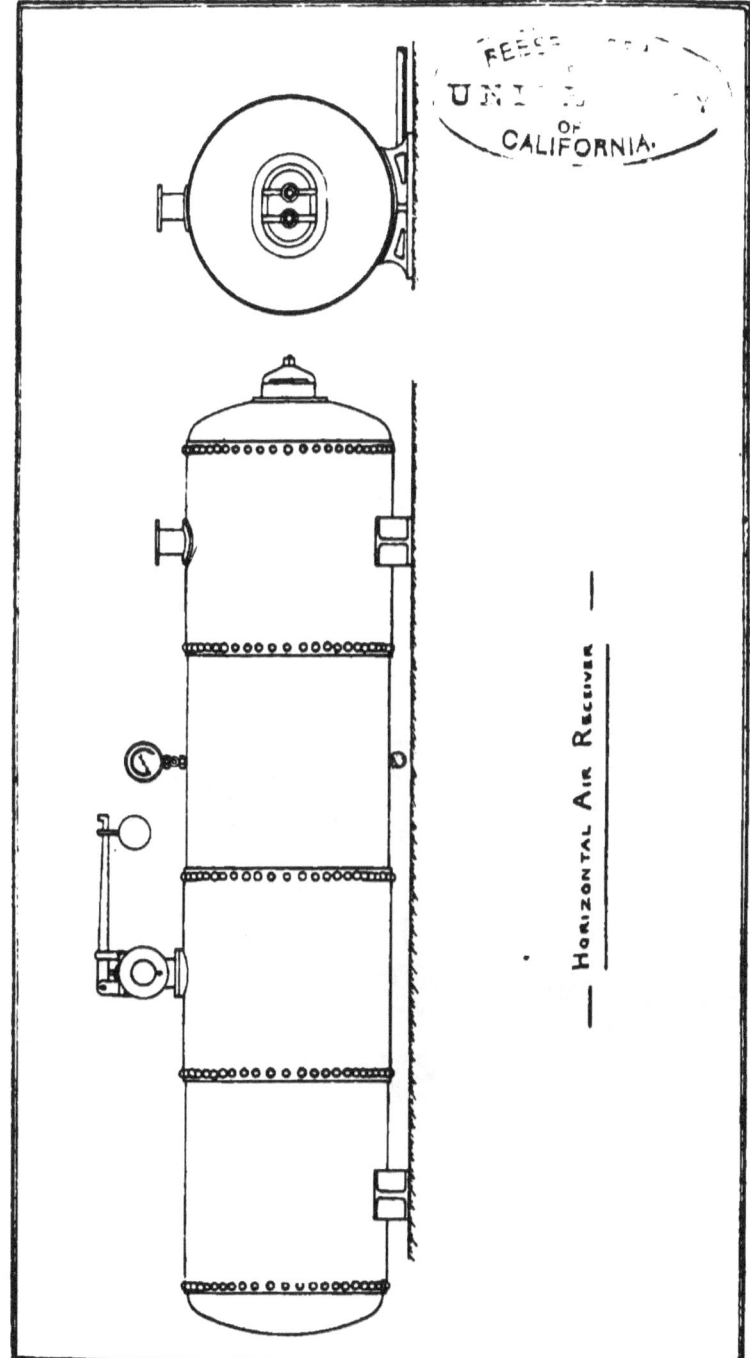

FIG. 71.— Horizontal Air Receiver.

SPECIAL BLACKSMITH TOOLS FOR DRILL BITS.

Fig. 72.	Fig. 73.	Fig. 74.	Fig. 75.	Fig. 76.
Sow.	Dolly.	Spreader.	Flatter.	Swedge.

RIX PATENT HOSE COUPLINGS.

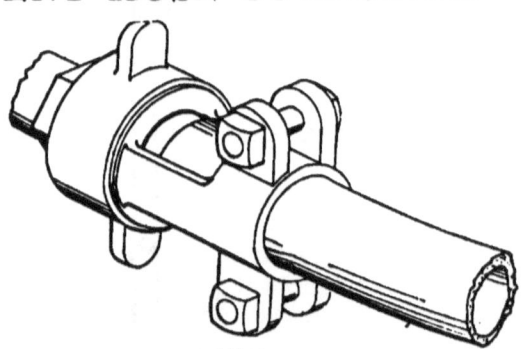

Fig. 77.

This coupling is the only coupling which will stay on a hose under all conditions of use. They have been used successfully at 600 lbs. per square inch, and are perfectly reliable. The nature of the coupling is such that it is rigidly connected to the hose, and nothing but the tearing away of the hose itself will separate it from the coupling.

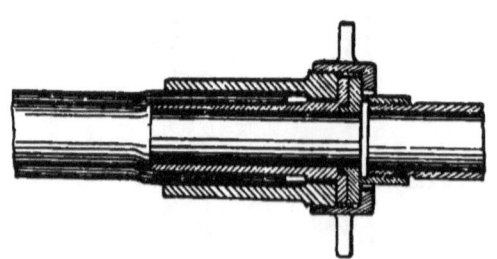

Fig. 78.

SIZES AS FOLLOWS:

For 1-inch 4 or 5 ply Hose.
For ¾-inch 4 or 5 ply Hose.
For 2-inch 4 or 5 ply Hose.

LUBRICATORS AND LUBRICANTS.

All of our Compressors for ordinary pressures, that is, up to 200 lbs. per square inch, are provided with the Ellis Sight Feed Lubricator for the air cylinders. This, or some similar device, is the only method for certain and economical lubrication. The ordinary oil cup delivers its entire contents in a short time, and there is no means of knowing when it requires filling, except by opening it. For all of our crank pins we use the Economy Oiler, which feeds only when the compressor is running. We have reports stating that one filling of one of these $4\frac{1}{2}$-ounce oilers on the crank of a 10-inch compressor lasted four weeks of continuous run.

The ordinary cylinder lubricating oils will not suffice for single stage dry compressor cylinders, where the compression is almost adiabatic from 200 degrees to 400 degrees, depending on the pressures. Poor oils are decomposed at these temperatures, and form combustible gases which may explode with dangerous effect. There was an explosion of this kind in the Idaho Mine, Grass Valley, a number of years ago, which destroyed several hundred feet of 6-inch air pipe in the shaft. Oils of at least 600 degrees fire test should be used. Any oil which burns on the outlet valves, leaving a hard, black, rubber-like substance, is not fit to be used.

Some engineers mix kerosene or coal oil with their cylinder lubricant, to cut the deposit and dirt from their valves, but it is a dangerous practise and will lead to accident, because the fire test of coal oil does not ordinarily exceed 175 degrees Fahr.

We carry in stock special oils for Compressors and Rock Drills, known as

RIX COMPRESSOR OIL.

RIX ROCK DRILL OIL.

ELLIS AIR CYLINDER OIL CUP.

The cylinders of Air Compressors are generally lubricated with a plain oil cup, and a great deal of difficulty is encountered in making this feed steady enough for practical purposes. Either the cup will not feed at all, or it will feed its entire contents in a few minutes. The lubricator which we are offering is a special lubricator, designed so that the pressure of air in the cylinder will force the oil through a small opening, which may be regulated, into the cylinder. The drops may be regulated as slow or as fast as necessary, and are made to drop in plain view, so as to make it a drop sight lubricator, something entirely new for air compressing cylinders and which we feel sure will be a great relief and satisfaction to those who have plants equipped with this class of machinery. It goes without saying that a lubricator of this kind will use about one-half of the oil that the ordinary lubricators require.

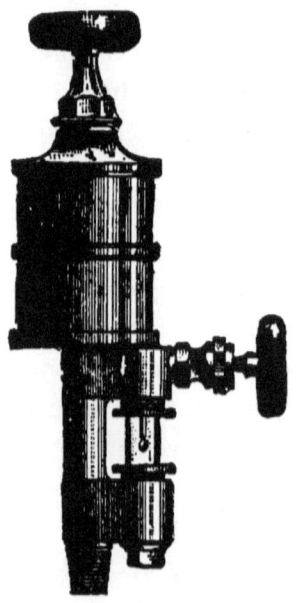

Fig. 79.

Made in either brass or nickel-plated finish, in the following sizes: ¼-pint, ⅓-pint, ½-pint, 1-pint, 1-quart.

APPENDIX.

USEFUL TABLES,
TO BE USED IN THE CALCULATION OF COMPRESSED AIR PROBLEMS.

The following tables and data in general will be found useful in the calculation of Compressed Air Problems. These tables have been taken from Kent's Hand Book, from The Pelton Water Wheel Company's catalogue, and from Carnegie Phipps & Co.'s catalogue, and we desire to express to the publishers of these volumes our thanks.

CIRCUMFERENCES AND AREAS OF CIRCLES
Advancing by Eighths.

Diam.	Circum.	Area.	Diam.	Circum.	Area.	Diam.	Circum.	Area.
1/64	.04909	.00019	2 3/8	7.4613	4.4301	6 1/8	19.242	29.465
1/32	.09818	.00077	7/16	7.6576	4.6664	1/4	19.635	30.680
3/64	.14726	.00173	1/2	7.8540	4.9087	3/8	20.028	31.919
1/16	.19635	.00307	9/16	8.0503	5.1572	1/2	20.420	33.183
3/32	.29452	.00690	5/8	8.2467	5.4119	5/8	20.813	34.472
1/8	.39270	.01227	11/16	8.4430	5.6727	3/4	21.206	35.785
5/32	.49087	.01917	3/4	8.6394	5.9396	7/8	21.598	37.122
3/16	.58905	.02761	13/16	8.8357	6.2126	7.	21.991	38.485
7/32	.68722	.03758	7/8	9.0321	6.4918	1/8	22.384	39.871
			15/16	9.2284	6.7771	1/4	22.776	41.282
1/4	.78540	.04909				3/8	23.169	42.718
9/32	.88357	.06213	3.	9.4248	7.0686	1/2	23.562	44.179
5/16	.98175	.07670	1/16	9.6211	7.3662	5/8	23.955	45.664
11/32	1.0799	.09281	1/8	9.8175	7.6699	3/4	24.347	47.173
3/8	1.1781	.11045	3/16	10.014	7.9798	7/8	24.740	48.707
13/32	1.2763	.12962	1/4	10.210	8.2958	8.	25.133	50.265
7/16	1.3744	.15033	5/16	10.407	8.6179	1/8	25.525	51.849
15/32	1.4726	.17257	3/8	10.603	8.9462	1/4	25.918	53.456
			7/16	10.799	9.2806	3/8	26.311	55.088
1/2	1.5708	.19635	1/2	10.996	9.6211	1/2	26.704	56.745
17/32	1.6690	.22166	9/16	11.192	9.9678	5/8	27.096	58.426
9/16	1.7671	.24850	5/8	11.388	10.321	3/4	27.489	60.132
19/32	1.8653	.27688	11/16	11.585	10.680	7/8	27.882	61.862
5/8	1.9635	.30680	3/4	11.781	11.045	9.	28.274	63.617
21/32	2.0617	.33824	13/16	11.977	11.416	1/8	28.667	65.397
11/16	2.1598	.37122	7/8	12.174	11.793	1/4	29.060	67.201
23/32	2.2580	.40574	15/16	12.370	12.177	3/8	29.452	69.029
			4.	12.566	12.566	1/2	29.845	70.882
3/4	2.3562	.44179	1/16	12.763	12.962	5/8	30.238	72.760
25/32	2.4544	.47937	1/8	12.959	13.364	3/4	30.631	74.662
13/16	2.5525	.51849	3/16	13.155	13.772	7/8	31.023	76.589
27/32	2.6507	.55914	1/4	13.352	14.186	10.	31.416	78.540
7/8	2.7489	.60132	5/16	13.548	14.607	1/8	31.809	80.516
29/32	2.8471	.64504	3/8	13.744	15.033	1/4	32.201	82.516
15/16	2.9452	.69029	7/16	13.941	15.466	3/8	32.594	84.541
31/32	3.0434	.73708	1/2	14.137	15.904	1/2	32.987	86.590
			9/16	14.334	16.349	5/8	33.379	88.664
1.	3.1416	.7854	5/8	14.530	16.800	3/4	33.772	90.763
1/16	3.3379	.8866	11/16	14.726	17.257	7/8	34.165	92.886
1/8	3.5343	.9940	3/4	14.923	17.728	11.	34.558	95.033
3/16	3.7306	1.1075	13/16	15.119	18.190	1/8	34.950	97.205
1/4	3.9270	1.2272	7/8	15.315	18.665	1/4	35.343	99.402
5/16	4.1233	1.3530	15/16	15.512	19.147	3/8	35.736	101.62
3/8	4.3197	1.4849	5.	15.708	19.635	1/2	36.128	103.87
7/16	4.5160	1.6230	1/16	15.904	20.129	5/8	36.521	106.14
1/2	4.7124	1.7671	1/8	16.101	20.629	3/4	36.914	108.43
9/16	4.9087	1.9175	3/16	16.297	21.135	7/8	37.306	110.75
5/8	5.1051	2.0739	1/4	16.493	21.648	12.	37.699	113.10
11/16	5.3014	2.2365	5/16	16.690	22.166	1/8	38.092	115.47
3/4	5.4978	2.4053	3/8	16.886	22.691	1/4	38.485	117.86
13/16	5.6941	2.5802	7/16	17.082	23.221	3/8	38.877	120.28
7/8	5.8905	2.7612	1/2	17.279	23.758	1/2	39.270	122.72
15/16	6.0868	2.9483	9/16	17.475	24.301	5/8	39.663	125.19
			5/8	17.671	24.850	3/4	40.055	127.68
2.	6.2832	3.1416	11/16	17.868	25.406	7/8	40.448	130.19
1/16	6.4795	3.3410	3/4	18.064	25.967	13.	40.841	132.73
1/8	6.6759	3.5466	13-16	18.261	26.535	1/8	41.233	135.30
3/16	6.8722	3.7583	7/8	18.457	27.109	1/4	41.626	137.89
1/4	7.0686	3.9761	15-16	18.653	27.688	3/8	42.019	140.50
5/16	7.2649	4.2000	6.	18.850	28.274	1/2	42.412	143.14

Diam.	Circum.	Area	Diam.	Circum.	Area	Diam.	Circum.	Area
13 5/8	42.804	145.80	21 7/8	68.722	375.83	30 1/8	94.640	712.76
3/4	43.197	148.49	22	69.115	380.13	1/4	95.033	718.69
7/8	43.590	151.20	1/8	69.508	384.46	3/8	95.426	724.64
14	43.982	153.94	1/4	69.900	388.82	1/2	95.819	730.62
1/8	44.375	156.70	3/8	70.293	393.20	5/8	96.211	736.62
1/4	44.768	159.48	1/2	70.686	397.61	3/4	96.604	742.64
3/8	45.160	162.30	5/8	71.079	402.04	7/8	96.997	748.69
1/2	45.553	165.13	3/4	71.471	406.49	31	97.389	754.77
5/8	45.946	167.99	7/8	71.864	410.97	1/8	97.782	760.87
3/4	46.338	170.87	23	72.257	415.48	1/4	98.175	766.99
7/8	46.731	173.78	1/8	72.649	420.00	3/8	98.567	773.14
15	47.124	176.71	1/4	73.042	424.56	1/2	98.960	779.31
1/8	47.517	179.67	3/8	73.435	429.13	5/8	99.353	785.51
1/4	47.909	182.65	1/2	73.827	433.74	3/4	99.746	791.73
3/8	48.302	185.66	5/8	74.220	438.36	7/8	100.138	797.98
1/2	48.695	188.69	3/4	74.613	443.01	32	100.531	804.25
5/8	49.087	191.75	7/8	75.006	447.69	1/8	100.924	810.54
3/4	49.480	194.83	24	75.398	452.39	1/4	101.316	816.86
7/8	49.873	197.93	1/8	75.791	457.11	3/8	101.709	823.21
16	50.265	201.06	1/4	76.184	461.86	1/2	102.102	829.58
1/8	50.658	204.22	3/8	76.576	466.64	5/8	102.494	835.97
1/4	51.051	207.39	1/2	76.969	471.44	3/4	102.887	842.39
3/8	51.444	210.60	5/8	77.362	476.26	7/8	103.280	848.83
1/2	51.836	213.82	3/4	77.754	481.11	33	103.673	855.30
5/8	52.229	217.08	7/8	78.147	485.98	1/8	104.065	861.79
3/4	52.622	220.35	25	78.540	490.87	1/4	104.458	868.31
7/8	53.014	223.65	1/8	78.933	495.79	3/8	104.851	874.85
17	53.407	226.98	1/4	79.325	500.74	1/2	105.243	881.41
1/8	53.800	230.33	3/8	79.718	505.71	5/8	105.636	888.00
1/4	54.192	233.71	1/2	80.111	510.71	3/4	106.029	894.62
3/8	54.585	237.10	5/8	80.503	515.72	7/8	106.421	901.26
1/2	54.978	240.53	3/4	80.896	520.77	34	106.814	907.92
5/8	55.371	243.98	7/8	81.289	525.84	1/8	107.207	914.61
3/4	55.763	247.45	26	81.681	530.93	1/4	107.600	921.32
7/8	56.156	250.95	1/8	82.074	536.05	3/8	107.992	928.06
18	56.549	254.47	1/4	82.467	541.19	1/2	108.385	934.82
1/8	56.941	258.02	3/8	82.860	546.35	5/8	108.778	941.61
1/4	57.334	261.59	1/2	83.252	551.55	3/4	109.170	948.42
3/8	57.727	265.18	5/8	83.645	556.76	7/8	109.563	955.25
1/2	58.119	268.80	3/4	84.038	562.00	35	109.956	962.11
5/8	58.512	272.45	7/8	84.430	567.27	1/8	110.348	969.00
3/4	58.905	276.12	27	84.823	572.56	1/4	110.741	975.91
7/8	59.298	279.81	1/8	85.216	577.87	3/8	111.134	982.84
19	59.690	283.53	1/4	85.608	583.21	1/2	111.527	989.80
1/8	60.083	287.27	3/8	86.001	588.57	5/8	111.919	996.78
1/4	60.476	291.04	1/2	86.394	593.96	3/4	112.312	1003.8
3/8	60.868	294.83	5/8	86.786	599.37	7/8	112.705	1010.8
1/2	61.261	298.65	3/4	87.179	604.81	36	113.097	1017.9
5/8	61.654	302.49	7/8	87.572	610.27	1/8	113.490	1025.0
3/4	62.046	306.35	28	87.965	615.75	1/4	113.883	1032.1
7/8	62.439	310.24	1/8	88.357	621.26	3/8	114.275	1039.2
20	62.832	314.16	1/4	88.750	626.80	1/2	114.668	1046.3
1/8	63.225	318.10	3/8	89.143	632.36	5/8	115.061	1053.5
1/4	63.617	322.06	1/2	89.535	637.94	3/4	115.454	1060.7
3/8	64.010	326.05	5/8	89.928	643.55	7/8	115.846	1068.0
1/2	64.403	330.06	3/4	90.321	649.18	37	116.239	1075.2
5/8	64.795	334.10	7/8	90.713	654.84	1/8	116.632	1082.5
3/4	65.188	338.16	29	91.106	660.52	1/4	117.024	1089.8
7/8	65.581	342.25	1/8	91.499	666.23	3/8	117.417	1097.1
21	65.973	346.36	1/4	91.892	671.96	1/2	117.810	1104.5
1/8	66.366	350.50	3/8	92.284	677.71	5/8	118.202	1111.8
1/4	66.759	354.66	1/2	92.677	683.49	3/4	118.596	1119.2
3/8	67.152	358.84	5/8	93.070	689.30	7/8	118.988	1126.7
1/2	67.544	363.05	3/4	93.462	695.13	38	119.381	1134.1
5/8	67.937	367.28	7/8	93.855	700.98	1/8	119.773	1141.6
3/4	68.330	371.54	30	94.248	706.86	1/4	120.166	1149.1

Diam.	Circum.	Area.	Diam.	Circum.	Area.	Diam.	Circum.	Area.
38 3/8	120.559	1156.6	46 5/8	146.477	1707.4	54 7/8	172.395	2365.0
1/2	120.951	1164.2	3/4	146.869	1716.5	55.	172.788	2375.8
5/8	121.344	1171.7	7/8	147.262	1725.7	1/8	173.180	2386.6
3/4	121.737	1179.3	47.	147.655	1734.9	1/4	173.573	2397.5
7/8	122.129	1186.9	1/8	148.048	1744.2	3/8	173.966	2408.3
39.	122.522	1194.6	1/4	148.440	1753.5	1/2	174.358	2419.2
1/8	122.915	1202.3	3/8	148.833	1762.7	5/8	174.751	2430.1
1/4	123.308	1210.0	1/2	149.226	1772.1	3/4	175.144	2441.1
3/8	123.700	1217.7	5/8	149.618	1781.4	7/8	175.536	2452.0
1/2	124.093	1225.4	3/4	150.011	1790.8	56.	175.929	2463.0
5/8	124.486	1233.2	7/8	150.404	1800.1	1/8	176.322	2474.0
3/4	124.878	1241.0	48.	150.796	1809.6	1/4	176.715	2485.0
7/8	125.271	1248.8	1/8	151.189	1819.0	3/8	177.107	2496.1
40.	125.664	1256.6	1/4	151.582	1828.5	1/2	177.500	2507.2
1/8	126.056	1264.5	3/8	151.975	1837.9	5/8	177.893	2518.3
1/4	126.449	1272.4	1/2	152.367	1847.5	3/4	178.285	2529.4
3/8	126.842	1280.3	5/8	152.760	1857.0	7/8	178.678	2540.6
1/2	127.235	1288.2	3/4	153.153	1866.5	57.	179.071	2551.8
5/8	127.627	1296.2	7/8	153.545	1876.1	1/8	179.463	2563.0
3/4	128.020	1304.2	49.	153.938	1885.7	1/4	179.856	2574.2
7/8	128.413	1312.2	1/8	154.331	1895.4	3/8	180.249	2585.4
41.	128.805	1320.3	1/4	154.723	1905.0	1/2	180.642	2596.7
1/8	129.198	1328.3	3/8	155.116	1914.7	5/8	181.034	2608.0
1/4	129.591	1336.4	1/2	155.509	1924.4	3/4	181.427	2619.4
3/8	129.983	1344.5	5/8	155.902	1934.2	7/8	181.820	2630.7
1/2	130.376	1352.7	3/4	156.294	1943.9	58.	182.212	2642.1
5/8	130.769	1360.8	7/8	156.687	1953.7	1/8	182.605	2653.5
3/4	131.161	1369.0	50.	157.080	1963.5	1/4	182.998	2664.9
7/8	131.554	1377.2	1/8	157.472	1973.3	3/8	183.390	2676.4
42.	131.947	1385.4	1/4	157.865	1983.2	1/2	183.783	2687.8
1/8	132.340	1393.7	3/8	158.258	1993.1	5/8	184.176	2699.3
1/4	132.732	1402.0	1/2	158.650	2003.0	3/4	184.569	2710.9
3/8	133.125	1410.3	5/8	159.043	2012.9	7/8	184.961	2722.4
1/2	133.518	1418.6	3/4	159.436	2022.8	59.	185.354	2734.0
5/8	133.910	1427.0	7/8	159.829	2032.8	1/8	185.747	2745.6
3/4	134.303	1435.4	51.	160.221	2042.8	1/4	186.139	2757.2
7/8	134.696	1443.8	1/8	160.614	2052.8	3/8	186.532	2768.8
43.	135.088	1452.2	1/4	161.007	2062.9	1/2	186.925	2780.5
1/8	135.481	1460.7	3/8	161.399	2073.0	5/8	187.317	2792.2
1/4	135.874	1469.1	1/2	161.792	2083.1	3/4	187.710	2803.9
3/8	136.267	1477.6	5/8	162.185	2093.2	7/8	188.103	2815.7
1/2	136.659	1486.2	3/4	162.577	2103.3	60.	188.496	2827.4
5/8	137.052	1494.7	7/8	162.970	2113.5	1/8	188.888	2839.2
3/4	137.445	1503.3	52.	163.363	2123.7	1/4	189.281	2851.0
7/8	137.837	1511.9	1/8	163.756	2133.9	3/8	189.674	2862.9
44.	138.230	1520.5	1/4	164.148	2144.2	1/2	190.066	2874.8
1/8	138.623	1529.2	3/8	164.541	2154.5	5/8	190.459	2886.6
1/4	139.015	1537.9	1/2	164.934	2164.8	3/4	190.852	2898.6
3/8	139.408	1546.6	5/8	165.326	2175.1	7/8	191.244	2910.5
1/2	139.801	1555.3	3/4	165.719	2185.4	61.	191.637	2922.5
5/8	140.194	1564.0	7/8	166.112	2195.8	1/8	192.030	2934.5
3/4	140.586	1572.8	53.	166.504	2206.2	1/4	192.423	2946.5
7/8	140.979	1581.6	1/8	166.897	2216.6	3/8	192.815	2958.5
45.	141.372	1590.4	1/4	167.290	2227.0	1/2	193.208	2970.6
1/8	141.764	1599.3	3/8	167.683	2237.5	5/8	193.601	2982.7
1/4	142.157	1608.2	1/2	168.075	2248.0	3/4	193.993	2994.8
3/8	142.550	1617.0	5/8	168.468	2258.5	7/8	194.386	3006.9
1/2	142.942	1626.0	3/4	168.861	2269.1	62.	194.779	3019.1
5/8	143.335	1634.9	7/8	169.253	2279.6	1/8	195.171	3031.3
3/4	143.728	1643.9	54.	169.646	2290.2	1/4	195.564	3043.5
7/8	144.121	1652.9	1/8	170.039	2300.8	3/8	195.957	3055.7
46.	144.513	1661.9	1/4	170.431	2311.5	1/2	196.350	3068.0
1/8	144.906	1670.9	3/8	170.824	2322.1	5/8	196.742	3080.3
1/4	145.299	1680.0	1/2	171.217	2332.8	3/4	197.135	3092.6
3/8	145.691	1689.1	5/8	171.609	2343.5	7/8	197.528	3104.9
1/2	146.084	1698.2	3/4	172.002	2354.3	63.	197.920	3117.2

FIFTH ROOTS AND FIFTH POWERS.
(Abridged from Trautwine.)

No. or Root	Power	No. or Root	Power	No. or Root	Power	No. or Root	Power	No. or Root	Power
.10	.000010	3.7	693.440	9.8	90392	21.8	4923597	40	102400000
.15	.000075	3.8	792.352	9.9	95099	22.0	5153632	41	115856201
.20	.000320	3.9	902.242	10.0	100000	22.2	5392186	42	130691232
.25	.000977	4.0	1024.00	10.2	110408	22.4	5639493	43	147008443
.30	.002430	4.1	1158.56	10.4	121665	22.6	5895793	44	164916224
.35	.005252	4.2	1306.91	10.6	133823	22.8	6161327	45	184528125
.40	.010240	4.3	1470.08	10.8	146933	23.0	6436343	46	205962976
.45	.018453	4.4	1649.16	11.0	161051	23.2	6721093	47	229345007
.50	.031250	4.5	1845.28	11.2	176234	23.4	7015834	48	254803968
.55	.050328	4.6	2059.63	11.4	192541	23.6	7320825	49	282475249
.60	.077760	4.7	2293.45	11.6	210034	23.8	7636332	50	312500000
.65	.116029	4.8	2548.04	11.8	228776	24.0	7962624	51	345025251
.70	.168070	4.9	2824.75	12.0	248832	24.2	8299976	52	380204032
.75	.237305	5.0	3125.00	12.2	270271	24.4	8648666	53	418195493
.80	.327680	5.1	3450.25	12.4	293163	24.6	9008978	54	459165024
.85	.443705	5.2	3802.04	12.6	317580	24.8	9381200	55	503284375
.90	.590490	5.3	4181.95	12.8	343597	25.0	9765625	56	550731776
.95	.773781	5.4	4591.65	13.0	371293	25.2	10162550	57	601692057
1.00	1.00000	5.5	5032.84	13.2	400746	25.4	10572278	58	656356768
1.05	1.27628	5.6	5507.32	13.4	432040	25.6	10995116	59	714924299
1.10	1.61051	5.7	6016.92	13.6	465259	25.8	11431377	60	777600000
1.15	2.01135	5.8	6563.57	13.8	500490	26.0	11881376	61	844596301
1.20	2.48832	5.9	7149.24	14.0	537824	26.2	12345437	62	916132832
1.25	3.05176	6.0	7776.00	14.2	577353	26.4	12823886	63	992436543
1.30	3.71293	6.1	8445.96	14.4	619174	26.6	13317055	64	1073741824
1.35	4.48403	6.2	9161.33	14.6	663383	26.8	13825281	65	1160290625
1.40	5.37824	6.3	9924.37	14.8	710082	27.0	14348907	66	1252332576
1.45	6.40973	6.4	10737	15.0	759375	27.2	14888280	67	1350125107
1.50	7.59375	6.5	11603	15.2	811368	27.4	15443752	68	1453933568
1.55	8.94661	6.6	12523	15.4	866171	27.6	16015681	69	1564031349
1.60	10.4858	6.7	13501	15.6	923896	27.8	16604430	70	1680700000
1.65	12.2298	6.8	14539	15.8	984658	28.0	17210368	71	1804299351
1.70	14.1986	6.9	15640	16.0	1048576	28.2	17833868	72	1934917632
1.75	16.3141	7.0	16807	16.2	1115771	28.4	18475309	73	2073071593
1.80	18.8957	7.1	18042	16.4	1186367	28.6	19135075	74	2219006624
1.85	21.6700	7.2	19349	16.6	1260493	28.8	19813557	75	2373046875
1.90	24.7610	7.3	20731	16.8	1338278	29.0	20511149	76	2535525376
1.95	28.1951	7.4	22190	17.0	1419857	29.2	21228253	77	2706784157
2.00	32.0000	7.5	23730	17.2	1505366	29.4	21965275	78	2887174368
2.05	36.2051	7.6	25355	17.4	1594947	29.6	22722628	79	3077056399
2.10	40.8410	7.7	27068	17.6	1688742	29.8	23500728	80	3276800000
2.15	45.9401	7.8	28872	17.8	1786899	30.0	24300000	81	3486784401
2.20	51.5363	7.9	30771	18.0	1889568	30.5	26393634	82	3707398432
2.25	57.6650	8.0	32768	18.2	1996903	31.0	28629151	83	3939040643
2.30	64.3634	8.1	34868	18.4	2109061	31.5	31013642	84	4182119424
2.35	71.6703	8.2	37074	18.6	2226203	32.0	33554432	85	4437053125
2.40	79.6262	8.3	39390	18.8	2348493	32.5	36259082	86	4704270176
2.45	88.2735	8.4	41821	19.0	2476099	33.0	39135393	87	4984209207
2.50	97.6562	8.5	44371	19.2	2609193	33.5	42191410	88	5277319168
2.55	107.820	8.6	47043	19.4	2747949	34.0	45435424	89	5584059449
2.60	118.814	8.7	49842	19.6	2892547	34.5	48875980	90	5904900000
2.70	143.489	8.8	52773	19.8	3043168	35.0	52521875	91	6240321451
2.80	172.104	8.9	55841	20.0	3200000	35.5	56382167	92	6590815232
2.90	205.111	9.0	59049	20.2	3363232	36.0	60466176	93	6956883693
3.00	243.000	9.1	62403	20.4	3533059	36.5	64783487	94	7339040224
3.10	286.292	9.2	65908	20.6	3709677	37.0	69343957	95	7737809375
3.20	335.544	9.3	69569	20.8	3893289	37.5	74157715	96	8153726976
3.30	391.354	9.4	73390	21.0	4084101	38.0	79235168	97	8587340257
3.40	454.354	9.5	77378	21.2	4282322	38.5	84587005	98	9039207968
3.50	525.219	9.6	81537	21.4	4488166	39.0	90224199	99	9509900499
3.60	604.662	9.7	85873	21.6	4701850	39.5	96158012		

SQUARES, CUBES AND RECIPROCALS.

Nos.	Squares.	Cubes.	Reciprocals.	Nos.	Squares.	Cubes.	Reciprocals.
1	1	1	1.000000000	51	26 01	132 651	.019607843
2	4	8	.500000000	52	27 04	140 608	.019230769
3	9	27	.333333333	53	28 09	148 877	.018867925
4	16	64	.250000000	54	29 16	157 464	.018518519
5	25	125	.200000000	55	30 25	166 375	.018181818
6	36	216	.166666667	56	31 36	175 616	.017857143
7	49	343	.142857143	57	32 49	185 193	.017543860
8	64	512	.125000000	58	33 64	195 112	.017241379
9	81	729	.111111111	59	34 81	205 379	.016949153
10	1 00	1 000	.100000000	60	36 00	216 000	.016666667
11	1 21	1 331	.090909091	61	37 21	226 981	.016393443
12	1 44	1 728	.083333333	62	38 44	238 328	.016129032
13	1 69	2 197	.076923077	63	39 69	250 047	.015873016
14	1 96	2 744	.071428571	64	40 96	262 144	.015625000
15	2 25	3 375	.066666667	65	42 25	274 625	.015384615
16	2 56	4 096	.062500000	66	43 56	287 496	.015151515
17	2 89	4 913	.058823529	67	44 89	300 763	.014925373
18	3 24	5 832	.055555556	68	46 24	314 432	.014705882
19	3 61	6 859	.052631579	69	47 61	328 509	.014492754
20	4 00	8 000	.050000000	70	49 00	343 000	.014285714
21	4 41	9 261	.047619048	71	50 41	357 911	.014081507
22	4 84	10 648	.045454545	72	51 84	373 248	.013888889
23	5 29	12 167	.043478260	73	53 29	389 017	.013698630
24	5 76	13 824	.041666667	74	54 76	405 224	.013513514
25	6 25	15 625	.040000000	75	56 25	421 875	.013333333
26	6 76	17 576	.038461538	76	57 76	438 976	.013157895
27	7 29	19 683	.037037037	77	59 29	456 533	.012987013
28	7 84	21 952	.035714286	78	60 84	474 552	.012820513
29	8 41	24 389	.034482759	79	62 41	493 039	.012658228
30	9 00	27 000	.033333333	80	64 00	512 000	.012500000
31	9 61	29 791	.032258065	81	65 61	531 441	.012345679
32	10 24	32 768	.031250000	82	67 24	551 368	.012195122
33	10 89	35 937	.030303030	83	68 89	571 787	.012048193
34	11 56	39 304	.029411765	84	70 56	592 704	.011904762
35	12 25	42 875	.028571429	85	72 25	614 125	.011764706
36	12 96	46 656	.027777778	86	73 96	636 056	.011627907
37	13 69	50 653	.027027027	87	75 69	658 503	.011494253
38	14 44	54 872	.026315789	88	77 44	681 472	.011363636
39	15 21	59 319	.025641026	89	79 21	704 969	.011235955
40	16 00	64 000	.025000000	90	81 00	729 000	.011111111
41	16 81	68 921	.024390244	91	82 81	753 571	.010989011
42	17 64	74 088	.023809524	92	84 64	778 688	.010869565
43	18 49	79 507	.023255814	93	86 49	804 357	.010752688
44	19 36	85 184	.022727273	94	88 36	830 584	.010638298
45	20 25	91 125	.022222222	95	90 25	857 375	.010526316
46	21 16	97 336	.021739130	96	92 16	884 736	.010416667
47	22 09	103 823	.021276600	97	94 09	912 673	.010309278
48	23 04	110 592	.020833333	98	96 04	941 192	.010204082
49	24 01	117 649	.020408163	99	98 01	970 299	.010101010
50	25 00	125 000	.020000000	100	1 00 00	1 000 000	.010000000

SQUARES, CUBES AND RECIPROCALS—Continued.

Nos.	Squares.	Cubes.	Reciprocals.	Nos.	Squares.	Cubes.	Reciprocals.
101	1 02 01	1 030 301	.009900990	151	2 28 01	3 442 951	.006622517
102	1 04 04	1 061 208	.009803922	152	2 31 04	3 511 808	.006578947
103	1 06 09	1 092 727	.009708738	153	2 34 09	3 581 577	.006535948
104	1 08 16	1 124 864	.009615385	154	2 37 16	3 652 264	.006493506
105	1 10 25	1 157 625	.009523810	155	2 40 25	3 723 875	.006451613
106	1 12 36	1 191 016	.009433962	156	2 43 36	3 796 416	.006410256
107	1 14 49	1 225 043	.009345794	157	2 46 49	3 869 893	.006369427
108	1 16 64	1 259 712	.009259259	158	2 49 64	3 944 312	.006329114
109	1 18 81	1 295 029	.009174312	159	2 52 81	4 019 679	.006289308
110	1 21 00	1 331 000	.009090909	160	2 56 00	4 096 000	.006250000
111	1 23 21	1 367 631	.009009009	161	2 59 21	4 173 281	.006211180
112	1 25 44	1 404 928	.008928571	162	2 62 44	4 251 5 8	.006172840
113	1 27 69	1 442 897	.008849558	163	2 65 69	4 330 747	.006134969
114	1 29 96	1 481 544	.008771930	164	2 68 96	4 410 944	.006097561
115	1 32 25	1 520 875	.008695652	165	2 72 25	4 492 125	.006060606
116	1 34 56	1 560 896	.008620690	166	2 75 56	4 574 296	.006024096
117	1 36 89	1 601 613	.008547009	167	2 78 89	4 657 463	.005988024
118	1 39 24	1 643 032	.008474576	168	2 82 24	4 741 632	.005952381
119	1 41 61	1 685 159	.008403361	169	2 85 61	4 826 809	.005917160
120	1 44 00	1 728 000	.008333333	170	2 89 00	4 913 000	.005882353
121	1 46 41	1 771 561	.008264463	171	2 92 41	5 000 211	.005847953
122	1 48 84	1 815 848	.008196721	172	2 95 84	5 088 448	.005813953
123	1 51 29	1 860 867	.008130081	173	2 99 29	5 177 717	.005780347
124	1 53 76	1 906 624	.008064516	174	3 02 76	5 268 024	.005747126
125	1 56 25	1 953 125	.008000000	175	3 06 25	5 359 375	.005714286
126	1 58 76	2 000 376	.007936508	176	3 09 76	5 451 776	.005681818
127	1 61 29	2 048 383	.007874016	177	3 13 29	5 545 233	.005649718
128	1 63 84	2 097 152	.007812500	178	3 16 84	5 639 752	.005617978
129	1 66 41	2 146 689	.007751938	179	3 20 41	5 735 339	.005586592
130	1 69 00	2 197 000	.007692308	180	3 24 00	5 832 000	.005555556
131	1 71 61	2 248 091	.007633588	181	3 27 61	5 929 741	.005524862
132	1 74 24	2 299 968	.007575758	182	3 31 24	6 028 568	.005494505
133	1 76 89	2 352 637	.007518797	183	3 34 89	6 128 487	.005464481
134	1 79 56	2 406 104	.007462687	184	3 38 56	6 229 504	.005434783
135	1 82 25	2 460 375	.007407407	185	3 42 25	6 331 625	.005405405
136	1 84 96	2 515 456	.007352941	186	3 45 96	6 434 856	.005376344
137	1 87 69	2 571 353	.007299270	187	3 49 69	6 539 203	.005347594
138	1 90 44	2 628 072	.007246377	188	3 53 44	6 644 672	.005319149
139	1 93 21	2 685 619	.007194245	189	3 57 21	6 751 269	.005291005
140	1 96 00	2 744 000	.007142857	190	3 61 00	6 859 000	.005263158
141	1 98 81	2 803 221	.007092199	191	3 64 81	6 967 871	.005235602
142	2 01 64	2 863 288	.007042254	192	3 68 64	7 077 888	.005208333
143	2 04 49	2 924 207	.006993007	193	3 72 49	7 189 057	.005181347
144	2 07 36	2 985 984	.006944444	194	3 76 36	7 301 384	.005154639
145	2 10 25	3 048 625	.006896552	195	3 80 25	7 414 875	.005128205
146	2 13 16	3 112 136	.006849315	196	3 84 16	7 529 536	.005102041
147	2 16 09	3 176 523	.006802721	197	3 88 09	7 645 373	.005076142
148	2 19 04	3 241 792	.006756757	198	3 92 04	7 762 392	.005050505
149	2 22 01	3 307 949	.006711409	199	3 96 01	7 880 599	.005025126
150	2 25 00	3 375 000	.006666667	200	4 00 00	8 000 000	.005000000

SQUARES, CUBES AND RECIPROCALS—Continued.

Nos.	Squares.	Cubes.	Reciprocals.	Nos.	Squares.	Cubes.	Reciprocals.
201	4 04 01	8 120 601	.004975124	251	6 30 01	15 813 251	.003984064
202	4 08 04	8 242 408	.004950495	252	6 35 04	16 003 008	.003968254
203	4 12 09	8 365 427	.004926108	253	6 40 09	16 194 277	.003952569
204	4 16 16	8 489 664	.004901961	254	6 45 16	16 387 064	.003937008
205	4 20 25	8 615 125	.004878049	255	6 50 25	16 581 375	.003921569
206	4 24 36	8 741 816	.004854369	256	6 55 36	16 777 216	.003906250
207	4 28 49	8 869 743	.004830918	257	6 60 49	16 974 593	.003891051
208	4 32 64	8 998 912	.004807692	258	6 65 64	17 173 512	.003875969
209	4 36 81	9 129 329	.004784689	259	6 70 81	17 373 979	.003861004
210	4 41 00	9 261 000	.004761905	260	6 76 00	17 576 000	.003846154
211	4 45 21	9 393 931	.004739336	261	6 81 21	17 779 581	.003831418
212	4 49 44	9 528 128	.004716981	262	6 86 44	17 984 728	.003816794
213	4 53 69	9 663 597	.004694836	263	6 91 69	18 191 447	.003802281
214	4 57 96	9 800 344	.004672897	264	6 96 96	18 399 744	.003787879
215	4 62 25	9 938 375	.004651163	265	7 02 25	18 609 625	.003773585
216	4 66 56	10 077 696	.004629630	266	7 07 56	18 821 096	.003759398
217	4 70 89	10 218 313	.004608295	267	7 12 89	19 034 163	.003745318
218	4 75 24	10 360 232	.004587156	268	7 18 24	19 248 832	.003731343
219	4 79 61	10 503 459	.004566210	269	7 23 61	19 465 109	.003717472
220	4 84 00	10 648 000	.004545455	270	7 29 00	19 683 000	.003703704
221	4 88 41	10 793 861	.004524887	271	7 34 41	19 902 511	.003690037
222	4 92 84	10 941 048	.004504505	272	7 39 84	20 123 648	.003676471
223	4 97 29	11 089 567	.004484305	273	7 45 29	20 346 417	.003663004
224	5 01 76	11 239 424	.004464286	274	7 50 76	20 570 824	.003649635
225	5 06 25	11 390 625	.004444444	275	7 56 25	20 796 875	.003636364
226	5 10 76	11 543 176	.004424779	276	7 61 76	21 024 576	.003623188
227	5 15 29	11 697 083	.004405286	277	7 67 29	21 253 933	.003610108
228	5 19 84	11 852 352	.004385965	278	7 72 84	21 484 952	.003597122
229	5 24 41	12 008 989	.004366812	279	7 78 41	21 717 639	.003584229
230	5 29 00	12 167 000	.004347826	280	7 84 00	21 952 000	.003571429
231	5 33 61	12 326 391	.004329004	281	7 89 61	22 188 041	.003558719
232	5 38 24	12 487 168	.004310345	282	7 95 24	22 425 768	.003546099
233	5 42 89	12 649 337	.004291845	283	8 00 89	22 665 187	.003533569
234	5 47 56	12 812 904	.004273504	284	8 06 56	22 906 304	.003521127
235	5 52 25	12 977 875	.004255319	285	8 12 25	23 149 125	.003508772
236	5 56 96	13 144 256	.004237288	286	8 17 96	23 393 656	.003496503
237	5 61 69	13 312 053	.004219400	287	8 23 69	23 639 903	.003484321
238	5 66 44	13 481 272	.004201681	288	8 29 44	23 887 872	.003472222
239	5 71 21	13 651 919	.004184100	289	8 35 21	24 137 569	.003460208
240	5 76 00	13 824 000	.004166667	290	8 41 00	24 389 000	.003448276
241	5 80 81	13 997 521	.004149378	291	8 46 81	24 642 171	.003436426
242	5 85 64	14 172 488	.004132231	292	8 52 64	24 897 088	.003424658
243	5 90 49	14 348 907	.004115226	293	8 58 49	25 153 757	.003412969
244	5 95 36	14 526 784	.004098361	294	8 64 36	25 412 184	.003401361
245	6 00 25	14 706 125	.004081633	295	8 70 25	25 672 375	.003389831
246	6 05 16	14 886 936	.004065041	296	8 76 16	25 934 336	.003378378
247	6 10 09	15 069 223	.004048583	297	8 82 09	26 198 073	.003367003
248	6 15 04	15 252 992	.004032258	298	8 88 04	26 463 592	.003355705
249	6 20 01	15 438 249	.004016064	299	8 94 01	26 730 899	.003344482
250	6 25 00	15 625 000	.004000000	300	9 00 00	27 000 000	.003333333

SQUARES, CUBES AND RECIPROCALS—Continued.

Nos.	Squares	Cubes.	Reciprocals.	Nos.	Squares.	Cubes.	Reciprocals.
301	9 06 01	27 270 901	.003322259	351	12 32 01	43 243 551	.002849003
302	9 12 04	27 543 608	.003311258	352	12 39 04	43 614 208	.002840909
303	9 18 09	27 818 127	.003300330	353	12 46 09	43 986 977	.002832861
304	9 24 16	28 094 464	.003289474	354	12 53 16	44 361 864	.002824859
305	9 30 25	28 372 625	.003278689	355	12 60 25	44 738 875	.002816901
306	9 36 36	28 652 616	.003267974	356	12 67 36	45 118 016	.002808989
307	9 42 49	28 934 443	.003257329	357	12 74 49	45 499 293	.002801120
308	9 48 64	29 218 112	.003246753	358	12 81 64	45 882 712	.002793296
309	9 54 81	29 503 629	.003236246	359	12 88 81	46 268 279	.002785515
310	9 61 00	29 791 000	.003225806	360	12 96 00	46 656 000	.002777778
311	9 67 21	30 080 231	.003215434	361	13 03 21	47 045 881	.002770083
312	9 73 44	30 371 328	.003205128	362	13 10 44	47 437 928	.002762431
313	9 79 69	30 664 297	.003194888	363	13 17 69	47 832 147	.002754821
314	9 85 96	30 959 144	.003184713	364	13 24 96	48 228 544	.002747253
315	9 92 25	31 255 875	.003174603	365	13 32 25	48 627 125	.002739726
316	9 98 56	31 554 496	.003164557	366	13 39 56	49 027 896	.002732240
317	10 04 89	31 855 013	.003154574	367	13 46 89	49 430 863	.002724796
318	10 11 24	32 157 432	.003144654	368	13 54 24	49 836 032	.002717391
319	10 17 61	32 461 759	.003134796	369	13 61 61	50 243 409	.002710027
320	10 24 00	32 768 000	.003125000	370	13 69 00	50 653 000	.002702703
321	10 30 41	33 076 161	.003115265	371	13 76 41	51 064 811	.002695418
322	10 36 84	33 386 248	.003105590	372	13 83 84	51 478 848	.002688172
323	10 43 29	33 698 267	.003095975	373	13 91 29	51 895 117	.002680965
324	10 49 76	34 012 224	.003086420	374	13 98 76	52 313 624	.002673797
325	10 56 25	34 328 125	.003076923	375	14 06 25	52 734 375	.002666667
326	10 62 76	34 645 976	.003067485	376	14 13 76	53 157 376	.002659574
327	10 69 29	34 965 783	.003058104	377	14 21 29	53 582 633	.002652520
328	10 75 84	35 287 552	.003048780	378	14 28 84	54 010 152	.002645503
329	10 82 41	35 611 289	.003039514	379	14 36 41	54 439 939	.002638522
330	10 89 00	35 937 000	.003030303	380	14 44 00	54 872 000	.002631579
331	10 95 61	36 264 691	.003021148	381	14 51 61	55 306 341	.002624672
332	11 02 24	36 594 368	.003012048	382	14 59 24	55 742 968	.002617801
333	11 08 89	36 926 037	.003003003	383	14 66 89	56 181 887	.002610966
334	11 15 56	37 259 704	.002994012	384	14 74 56	56 623 104	.002604167
335	11 22 25	37 595 375	.002985075	385	14 82 25	57 066 625	.002597403
336	11 28 96	37 933 056	.002976190	386	14 89 96	57 512 456	.002590674
337	11 35 69	38 272 753	.002967359	387	14 97 69	57 960 603	.002583979
338	11 42 44	38 614 472	.002958580	388	15 05 44	58 411 072	.002577320
339	11 49 21	38 958 219	.002949853	389	15 13 21	58 863 869	.002570694
340	11 56 00	39 304 000	.002941176	390	15 21 00	59 319 000	.002564103
341	11 62 81	39 651 821	.002932551	391	15 28 81	59 776 471	.002557545
342	11 69 64	40 001 688	.002923977	392	15 36 64	60 236 288	.002551020
343	11 76 49	40 353 607	.002915452	393	15 44 49	60 698 457	.002544529
344	11 83 36	40 707 584	.002906977	394	15 52 36	61 162 984	.002538071
345	11 90 25	41 063 625	.002898551	395	15 60 25	61 629 875	.002531646
346	11 97 16	41 421 736	.002890173	396	15 68 16	62 099 136	.002525253
347	12 04 09	41 781 923	.002881844	397	15 76 09	62 570 773	.002518892
348	12 11 04	42 144 192	.002873563	398	15 84 04	63 044 792	.002512563
349	12 18 01	42 508 549	.002865330	399	15 92 01	63 521 199	.002506266
350	12 25 00	42 875 000	.002857143	400	16 00 00	64 000 000	.002500000

SQUARES, CUBES AND RECIPROCALS—Continued.

Nos.	Squares.	Cubes.	Reciprocals.	Nos.	Squares.	Cubes.	Reciprocals.
401	16 08 01	64 481 201	.002493766	451	20 34 01	91 733 851	.002217295
402	16 16 04	64 964 808	.002487562	452	20 43 04	92 345 408	.002212389
403	16 24 09	65 450 827	.002481390	453	20 52 09	92 959 677	.002207506
404	16 32 16	65 939 264	.002475248	454	20 61 16	93 576 664	.002202643
405	16 40 25	66 430 125	.002469136	455	20 70 25	94 196 375	.002197802
406	16 48 36	66 923 416	.002463054	456	20 79 36	94 818 816	.002192982
407	16 56 49	67 419 143	.002457002	457	20 88 49	95 443 993	.002188184
408	16 64 64	67 917 312	.002450980	458	20 97 64	96 071 912	.002183406
409	16 72 81	68 417 929	.002444988	459	21 06 81	96 702 579	.002178649
410	16 81 00	68 921 000	.002439024	460	21 16 00	97 336 000	.002173913
411	16 89 21	69 426 531	.002433090	461	21 25 21	97 972 181	.002169197
412	16 97 44	69 934 528	.002427184	462	21 34 44	98 611 128	.002164502
413	17 05 69	70 444 997	.002421308	463	21 43 69	99 252 847	.002159827
414	17 13 96	70 957 944	.002415459	464	21 52 96	99 897 344	.002155172
415	17 22 25	71 473 875	.002409639	465	21 62 25	100 544 625	.002150538
416	17 30 56	71 991 296	.002403846	466	21 71 56	101 194 696	.002145923
417	17 38 89	72 511 713	.002398082	467	21 80 89	101 847 563	.002141328
418	17 47 24	73 034 632	.002392344	468	21 90 24	102 503 232	.002136752
419	17 55 61	73 560 059	.002386635	469	21 99 61	103 161 709	.002132196
420	17 64 00	74 088 000	.002380952	470	22 09 00	103 823 000	.002127660
421	17 72 41	74 618 461	.002375297	471	22 18 41	104 487 111	.002123142
422	17 80 84	75 151 448	.002369668	472	22 27 84	105 154 048	.002118644
423	17 89 29	75 686 967	.002364066	473	22 37 29	105 823 817	.002114165
424	17 97 76	76 225 024	.002358491	474	22 46 76	106 496 424	.002109705
425	18 06 25	76 765 625	.002352941	475	22 56 25	107 171 875	.002105263
426	18 14 76	77 308 776	.002347418	476	22 65 76	107 850 176	.002100840
427	18 23 29	77 854 483	.002341920	477	22 75 29	108 531 333	.002096436
428	18 31 84	78 402 752	.002336449	478	22 84 84	109 215 352	.002092050
429	18 40 41	78 953 589	.002331002	479	22 94 41	109 902 239	.002087683
430	18 49 00	79 507 000	.002325581	480	23 04 00	110 592 000	.002083333
431	18 57 61	80 062 991	.002320186	481	23 13 61	111 284 641	.002079002
432	18 66 24	80 621 568	.002314815	482	23 23 24	111 980 168	.002074689
433	18 74 89	81 182 737	.002309469	483	23 32 89	112 678 587	.002070393
434	18 83 56	81 746 504	.002304147	484	23 42 56	113 379 904	.002066116
435	18 92 25	82 312 875	.002298851	485	23 52 25	114 084 125	.002061856
436	19 00 96	82 881 856	.002293578	486	23 61 96	114 791 256	.002057613
437	19 09 69	83 453 453	.002288330	487	23 71 69	115 501 303	.002053388
438	19 18 44	84 027 672	.002283105	488	23 81 44	116 214 272	.002049180
439	19 27 21	84 604 519	.002277904	489	23 91 21	116 930 169	.002044990
440	19 36 00	85 184 000	.002272727	490	24 01 00	117 649 000	.002040816
441	19 44 81	85 766 121	.002267574	491	24 10 81	118 370 771	.002036660
442	19 53 64	86 350 888	.002262443	492	24 20 64	119 095 488	.002032520
443	19 62 49	86 938 307	.002257336	493	24 30 49	119 823 157	.002028398
444	19 71 36	87 528 384	.002252252	494	24 40 36	120 553 784	.002024291
445	19 80 25	88 121 125	.002247191	495	24 50 25	121 287 375	.002020202
446	19 89 16	88 716 536	.002242152	496	24 60 16	122 023 936	.002016129
447	19 98 09	89 314 623	.002237136	497	24 70 09	122 763 473	.002012072
448	20 07 04	89 915 392	.002232143	498	24 80 04	123 505 992	.002008032
449	20 16 01	90 518 849	.002227171	499	24 90 01	124 251 499	.002004008
450	20 25 00	91 125 000	.002222222	500	25 00 00	125 000 000	.002000000

TEMPERATURES, CENTIGRADE — FAHRENHEIT.

C.	F.	C.	F.	C.	F.	C.	F.	C.	F.
−40	−40.	26	78.8	92	197.6	158	316.4	224	435.1
−39	−38.2	27	80.6	93	199.4	159	318.2	225	437.
−38	−36.4	28	82.4	94	201.2	160	320.	226	438.8
−37	−34.6	29	84.2	95	203.	161	321.8	227	440.6
−36	−32.8	30	86.	96	204.8	162	323.6	228	442.
−35	−31.	31	87.8	97	206.6	163	325.4	229	444.2
−34	−29.2	32	89.6	98	208.4	164	327.2	230	446.
−33	−27.4	33	91.4	99	210.2	165	329.	231	447.8
−32	−25.6	34	93.2	100	212.	166	330.8	232	449.6
−31	−23.8	35	95.	101	213.8	167	332.6	233	451.4
−30	−22.	36	96.8	102	215.6	168	334.4	234	453.2
−29	−20.2	37	98.6	103	217.4	169	336.2	235	455.
−28	−18.4	38	100.4	104	219.2	170	338.	236	456.8
−27	−16.6	39	102.2	105	221.	171	339.8	237	458.6
−26	−14.8	40	104.	106	222.8	172	341.6	238	460.
−25	−13.	41	105.8	107	224.6	173	343.4	239	462.2
−24	−11.2	42	107.6	108	226.4	174	345.2	240	464.
−23	−9.4	43	109.4	109	228.2	175	347.	241	465.8
−22	−7.6	44	111.2	110	230.	176	348.8	242	467.6
−21	−5.8	45	113.	111	231.8	177	350.6	243	469.
−20	−4.	46	114.8	112	233.6	178	352.4	244	471.2
−19	−2.2	47	116.6	113	235.4	179	354.2	245	473.
−18	−0.4	48	118.4	114	237.2	180	356.	246	474.8
−17	+1.4	49	120.2	115	239.	181	357.8	247	476.6
−16	3.2	50	122.	116	240.8	182	359.6	248	478.4
−15	5.	51	123.8	117	242.6	183	361.4	249	480.2
−14	6.8	52	125.6	118	244.4	184	363.2	250	482.
−13	8.6	53	127.4	119	246.2	185	365.	251	483.8
−12	10.4	54	129.2	120	248.	186	366.8	252	485.
−11	12.2	55	131.	121	249.8	187	368.6	253	487.
−10	14.	56	132.8	122	251.6	188	370.4	254	489.
−9	15.8	57	134.6	123	253.4	189	372.2	255	491.
−8	17.6	58	136.4	124	255.2	190	374.	256	492.
−7	19.4	59	138.2	125	257.	191	375.8	257	494.
−6	21.2	60	140.	126	258.8	192	377.6	258	496.
−5	23.	61	141.8	127	260.6	193	379.4	259	498.
−4	24.8	62	143.6	128	262.4	194	381.2	260	500.
−3	26.6	63	145.4	129	264.2	195	383.	261	501.
−2	28.4	64	147.2	130	266.	196	384.8	262	503.
−1	30.2	65	149.	131	267.8	197	386.6	263	505.
0	32.	66	150.8	132	269.6	198	388.4	264	507.
+1	33.8	67	152.6	133	271.4	199	390.2	265	509.
2	35.6	68	154.4	134	273.2	200	392.	266	510.
3	37.4	69	156.2	135	275.	201	393.8	267	512.
4	39.2	70	158.	136	276.8	202	395.6	268	514.
5	41.	71	159.8	137	278.6	203	397.4	269	516.
6	42.8	72	161.6	138	280.4	204	399.2	270	518.
7	44.6	73	163.4	139	282.2	205	401.	271	519.
8	46.4	74	165.2	140	284.	206	402.8	272	521.
9	48.2	75	167.	141	285.8	207	404.6	273	523.
10	50.	76	168.8	142	287.6	208	406.4	274	525.
11	51.8	77	170.6	143	289.4	209	408.2	275	527.
12	53.6	78	172.4	144	291.2	210	410.	276	529.
13	55.4	79	174.2	145	293.	211	411.8	277	530.
14	57.2	80	176.	146	294.8	212	413.6	278	532.
15	59.	81	177.8	147	296.6	213	415.4	279	534.
16	60.8	82	179.6	148	298.4	214	417.2	280	536.
17	62.6	83	181.4	149	300.2	215	419.	281	537.
18	64.4	84	183.2	150	302.	216	420.8	282	539.
19	66.2	85	185.	151	303.8	217	422.6	283	541.
20	68.	86	186.8	152	305.6	218	424.4	284	543.
21	69.8	87	188.6	153	307.4	219	426.2	285	545.
22	71.6	88	190.4	154	309.2	220	428.	286	546.
23	73.4	89	192.2	155	311.	221	429.8	287	548.
24	75.2	90	194.	156	312.9	222	431.6	288	550.
25	77.	91	195.8	157	314.6	223	433.4	289	552.

TEMPERATURES, FAHRENHEIT AND CENTIGRADE.

F.	C.	F.	C.	F	C.	F	C.	F.	C.	F.	C.	F.	C.
−40	−40.	26	− 3.3	92	33.3	158	70.	224	106.7	290	143.3	360	182.2
−39	−39.4	27	− 2.8	93	33.9	159	70.6	225	107.2	291	143.9	370	187.8
−38	−38.9	28	− 2.2	94	34.4	160	71.1	226	107.8	292	144.4	380	193.3
−37	−38.3	29	− 1.7	95	35.	161	71.7	227	108.3	293	145.	390	198.9
−36	−37.8	30	− 1.1	96	35.6	162	72.2	228	108.9	294	145.6	400	204.4
−35	−37.2	31	− 0.6	97	36.1	163	72.8	229	109.4	295	146.1	410	210.
−34	−36.7	32	0.	98	36.7	164	73.3	230	110.	296	146.7	420	215.6
−33	−36.1	33	+ 0.6	99	37.2	165	73.9	231	110.6	297	147.2	430	221.1
−32	−35.6	34	1.1	100	37.8	166	74.4	232	111.1	298	147.8	440	226.7
−31	−35.	35	1.7	101	38.3	167	75.	233	111.7	299	148.3	450	232.2
−30	−34.4	36	2.2	102	38.9	168	75.6	234	112.2	300	148.9	460	237.8
−29	−33.9	37	2.8	103	39.4	169	76.1	235	112.8	301	149.4	470	243.3
−28	−33.3	38	3.3	104	40.	170	76.7	236	113.3	302	150.	480	248.9
−27	−32.8	39	3.9	105	40.6	171	77.2	237	113.9	303	150.6	490	254.4
−26	−32.2	40	4.4	106	41.1	172	77.8	238	114.4	304	151.1	500	260.
−25	−31.7	41	5.	107	41.7	173	78.3	239	115.	305	151.7	510	265.6
−24	−31.1	42	5.6	108	42.2	174	78.9	240	115.6	306	152.2	520	271.1
−23	−30.6	43	6.1	109	42.8	175	79.4	241	116.1	307	152.8	530	276.7
−22	−30.	44	6.7	110	43.3	176	80.	242	116.7	308	153.3	540	282.2
−21	−29.4	45	7.2	111	43.9	177	80.6	243	117.2	309	153.9	550	287.8
−20	−28.9	46	7.8	112	44.4	178	81.1	244	117.8	310	154.4	560	293.3
−19	−28.3	47	8.3	113	45.	179	81.7	245	118.3	311	155.	570	298.9
−18	−27.8	48	8.9	114	45.6	180	82.2	246	118.9	312	155.6	580	304.4
−17	−27.2	49	9.4	115	46.1	181	82.8	247	119.4	313	156.1	590	310.
−16	−26.7	50	10.	116	46.7	182	83.3	248	120.	314	156.7	600	315.6
−15	−26.1	51	10.6	117	47.2	183	83.9	249	120.6	315	157.2	610	321.1
−14	−25.6	52	11.1	118	47.8	184	84.4	250	121.1	316	157.8	620	326.7
−13	−25.	53	11.7	119	48.3	185	85.	251	121.7	317	158.3	630	332.2
−12	−24.4	54	12.2	120	48.9	186	85.6	252	122.2	318	158.9	640	337.8
−11	−23.9	55	12.8	121	49.4	187	86.1	253	122.8	319	159.4	650	343.3
−10	−23.3	56	13.3	122	50.	188	86.7	254	123.3	320	160.	660	348.9
− 9	−22.8	57	13.9	123	50.6	189	87.2	255	123.9	321	160.6	670	354.4
− 8	−22.2	58	14.4	124	51.1	190	87.8	256	124.4	322	161.1	680	360.
− 7	−21.7	59	15.	125	51.7	191	88.3	257	125.	323	161.7	690	365.6
− 6	−21.1	60	15.6	126	52.2	192	88.9	258	125.6	324	162.2	700	371.1
− 5	−20.6	61	16.1	127	52.8	193	89.4	259	126.1	325	162.8	710	376.7
− 4	−20.	62	16.7	128	53.3	194	90.	260	126.7	326	163.3	720	382.2
− 3	−19.4	63	17.2	129	53.9	195	90.6	261	127.2	327	163.9	730	387.8
− 2	−18.9	64	17.8	130	54.4	196	91.1	262	127.8	328	164.4	740	393.3
− 1	−18.3	65	18.3	131	55.	197	91.7	263	128.3	329	165.	750	398.9
0	−17.8	66	18.9	132	55.6	198	92.2	264	128.9	330	165.6	760	404.4
+1	−17.2	67	19.4	133	56.1	199	92.8	265	129.4	331	166.1	770	410.
2	−16.7	68	20.	134	56.7	200	93.3	266	130.	332	166.7	780	415.6
3	−16.1	69	20.6	135	57.2	201	93.9	267	130.6	333	167.2	790	421.1
4	−15.6	70	21.1	136	57.8	202	94.4	268	131.1	334	167.8	800	426.7
5	−15.	71	21.7	137	58.3	203	95.	269	131.7	335	168.3	810	432.2
6	−14.4	72	22.2	138	58.9	204	95.6	270	132.2	336	168.9	820	437.8
7	−13.9	73	22.8	139	59.4	205	96.1	271	132.8	337	169.4	830	443.3
8	−13.3	74	23.3	140	60.	206	96.7	272	133.3	338	170.	840	448.9
9	−12.8	75	23.9	141	60.6	207	97.2	273	133.9	339	170.6	850	454.4
10	−12.2	76	24.4	142	61.1	208	97.8	274	134.4	340	171.1	860	460.
11	−11.7	77	25.	143	61.7	209	98.3	275	135.	341	171.7	870	465.6
12	−11.1	78	25.6	144	62.2	210	98.9	276	135.6	342	172.2	880	471.1
13	−10.6	79	26.1	145	62.8	211	99.4	277	136.1	343	172.8	890	476.7
14	−10.	80	26.7	146	63.3	212	100.	278	136.7	344	173.3	900	482.2
15	− 9.4	81	27.2	147	63.9	213	100.6	279	137.2	345	173.9	910	487.8
16	− 8.9	82	27.8	148	64.4	214	101.1	280	137.8	346	174.4	920	493.3
17	− 8.3	83	28.3	149	65.	215	101.7	281	138.3	347	175.	930	498.9
18	− 7.8	84	28.9	150	65.6	216	102.2	282	138.9	348	175.6	940	504.4
19	− 7.2	85	29.4	151	66.1	217	102.8	283	139.4	349	176.1	950	510.
20	− 6.7	86	30.	152	66.7	218	103.3	284	140.	350	176.7	960	515.6
21	− 6.1	87	30.6	153	67.2	219	103.9	285	140.6	351	177.2	970	521.1
22	− 5.6	88	31.1	154	67.8	220	104.4	286	141.1	352	177.8	980	526.7
23	− 5.	89	31.7	155	68.3	221	105.	287	141.7	353	178.3	990	532.2
24	− 4.4	90	32.2	156	68.9	222	105.6	288	142.2	354	178.9	1000	537.8
25	− 3.9	91	32.8	157	69.4	223	106.1	289	142.8	355	179.4	1010	543.3

DECIMALS OF A FOOT FOR EACH 1/16 OF AN INCH.

Inch.	0"	1"	2"	3"	4"	5"
0	0	.0833	.1667	.2500	.3333	.4167
1/64	.0013	.0846	.1680	.2513	.3346	.4180
1/32	.0026	.0859	.1693	.2526	.3359	.4193
3/64	.0039	.0872	.1706	.2539	.3372	.4206
1/16	.0052	.0885	.1719	.2552	.3385	.4219
5/64	.0065	.0898	.1732	.2565	.3398	.4232
3/32	.0078	.0911	.1745	.2578	.3411	.4245
7/64	.0091	.0924	.1758	.2591	.3424	.4258
1/8	.0104	.0937	.1771	.2604	.3437	.4271
9/64	.0117	.0951	.1784	.2617	.3451	.4284
5/32	.0130	.0964	.1797	.2630	.3464	.4297
11/64	.0143	.0977	.1810	.2643	.3477	.4310
3/16	.0156	.0990	.1823	.2656	.3490	.4323
13/64	.0169	.1003	.1836	.2669	.3503	.4336
7/32	.0182	.1016	.1849	.2682	.3516	.4349
15/64	.0195	.1029	.1862	.2695	.3529	.4362
1/4	.0208	.1042	.1875	.2708	.3542	.4375
17/64	.0221	.1055	.1888	.2721	.3555	.4388
9/32	.0234	.1068	.1901	.2734	.3568	.4401
19/64	.0247	.1081	.1914	.2747	.3581	.4414
5/16	.0260	.1094	.1927	.2760	.3594	.4427
21/64	.0273	.1107	.1940	.2773	.3607	.4440
11/32	.0286	.1120	.1953	.2786	.3620	.4453
23/64	.0299	.1133	.1966	.2799	.3633	.4466
3/8	.0312	.1146	.1979	.2812	.3646	.4479
25/64	.0326	.1159	.1992	.2826	.3659	.4492
13/32	.0339	.1172	.2005	.2839	.3672	.4505
27/64	.0352	.1185	.2018	.2852	.3685	.4518
7/16	.0365	.1198	.2031	.2865	.3698	.4531
29/64	.0378	.1211	.2044	.2878	.3711	.4544
15/32	.0391	.1224	.2057	.2891	.3724	.4557
31/64	.0404	.1237	.2070	.2904	.3737	.4570
1/2	.0417	.1250	.2083	.2917	.3750	.4583

DECIMALS OF A FOOT FOR EACH 1/16 OF AN INCH

Inch.	6"	7"	8"	9"	10"	11"
0	.5000	.5833	.6667	.7500	.8333	.9167
1/16	.5013	.5846	.6680	.7513	.8346	.9180
1/8	.5026	.5859	.6693	.7526	.8359	.9193
3/16	.5039	.5872	.6706	.7539	.8372	.9206
1/4	.5052	.5885	.6719	.7552	.8385	.9219
5/16	.5065	.5898	.6732	.7565	.8398	.9232
3/8	.5078	.5911	.6745	.7578	.8411	.9245
7/16	.5091	.5924	.6758	.7591	.8424	.9258
1/2	.5104	.5937	.6771	.7604	.8437	.9271
9/16	.5117	.5951	.6784	.7617	.8451	.9284
5/8	.5130	.5964	.6797	.7630	.8464	.9297
11/16	.5143	.5977	.6810	.7643	.8477	.9310
3/4	.5156	.5990	.6823	.7656	.8490	.9323
13/16	.5169	.6003	.6836	.7669	.8503	.9336
7/8	.5182	.6016	.6849	.7682	.8516	.9349
15/16	.5195	.6029	.6862	.7695	.8529	.9362
1	.5208	.6042	.6875	.7708	.8542	.9375
1 1/16	.5221	.6055	.6888	.7721	.8555	.9388
1 1/8	.5234	.6068	.6901	.7734	.8568	.9401
1 3/16	.5247	.6081	.6914	.7747	.8581	.9414
1 1/4	.5260	.6094	.6927	.7760	.8594	.9427
1 5/16	.5273	.6107	.6940	.7773	.8607	.9440
1 3/8	.5286	.6120	.6953	.7786	.8620	.9453
1 7/16	.5299	.6133	.6966	.7799	.8633	.9466
1 1/2	.5312	.6146	.6979	.7812	.8646	.9479
1 9/16	.5326	.6159	.6992	.7826	.8659	.9492
1 5/8	.5339	.6172	.7005	.7839	.8672	.9505
1 11/16	.5352	.6185	.7018	.7852	.8685	.9518
1 3/4	.5365	.6198	.7031	.7865	.8698	.9531
1 13/16	.5378	.6211	.7044	.7878	.8711	.9544
1 7/8	.5391	.6224	.7057	.7891	.8724	.9557
1 15/16	.5404	.6237	.7070	.7904	.8737	.9570
2	.5417	.6250	.7083	.7917	.8750	.9583

DECIMALS OF A FOOT FOR EACH 1/64 OF AN INCH.

Inch.	0	1″	2″	3″	4″	5″
1/64	.0430	.1263	.2096	.2930	.3763	.4596
2/64	.0443	.1276	.2109	.2943	.3776	.4609
3/64	.0456	.1289	.2122	.2956	.3789	.4622
4/64	.0469	.1302	.2135	.2969	.3802	.4635
5/64	.0482	.1315	.2148	.2982	.3815	.4648
6/64	.0495	.1328	.2161	.2995	.3828	.4661
7/64	.0508	.1341	.2174	.3008	.3841	.4674
8/64	.0521	.1354	.2188	.3021	.3854	.4688
9/64	.0534	.1367	.2201	.3034	.3867	.4701
10/64	.0547	.1380	.2214	.3047	.3880	.4714
11/64	.0560	.1393	.2227	.3060	.3893	.4727
12/64	.0573	.1406	.2240	.3073	.3906	.4740
13/64	.0586	.1419	.2253	.3086	.3919	.4753
14/64	.0599	.1432	.2266	.3099	.3932	.4766
15/64	.0612	.1445	.2279	.3112	.3945	.4779
16/64	.0625	.1458	.2292	.3125	.3958	.4792
17/64	.0638	.1471	.2305	.3138	.3971	.4805
18/64	.0651	.1484	.2318	.3151	.3984	.4818
19/64	.0664	.1497	.2331	.3164	.3997	.4831
20/64	.0677	.1510	.2344	.3177	.4010	.4844
21/64	.0690	.1523	.2357	.3190	.4023	.4857
22/64	.0703	.1536	.2370	.3203	.4036	.4870
23/64	.0716	.1549	.2383	.3216	.4049	.4883
24/64	.0729	.1562	.2396	.3229	.4062	.4896
25/64	.0742	.1576	.2409	.3242	.4076	.4909
26/64	.0755	.1589	.2422	.3255	.4089	.4922
27/64	.0768	.1602	.2435	.3268	.4102	.4935
28/64	.0781	.1615	.2448	.3281	.4115	.4948
29/64	.0794	.1628	.2461	.3294	.4128	.4961
30/64	.0807	.1641	.2474	.3307	.4141	.4974
31/64	.0820	.1654	.2487	.3320	.4154	.4987

DECIMALS OF A FOOT FOR EACH 1/16 OF AN INCH.

Inch.	6″	7″	8″	9″	10″	11″
1/16	.5430	.6263	.7096	.7930	.8763	.9596
1/8	.5443	.6276	.7109	.7943	.8776	.9609
3/16	.5456	.6289	.7122	.7956	.8789	.9622
1/4	.5469	.6302	.7135	.7969	.8802	.9635
5/16	.5482	.6315	.7148	.7982	.8815	.9648
3/8	.5495	.6328	.7161	.7995	.8828	.9661
7/16	.5508	.6341	.7174	.8008	.8841	.9674
1/2	.5521	.6354	.7188	.8021	.8854	.9688
9/16	.5534	.6367	.7201	.8034	.8867	.9701
5/8	.5547	.6380	.7214	.8047	.8880	.9714
11/16	.5560	.6393	.7227	.8060	.8893	.9727
3/4	.5573	.6406	.7240	.8073	.8906	.9740
13/16	.5586	.6419	.7253	.8086	.8919	.9753
7/8	.5599	.6432	.7266	.8099	.8932	.9766
15/16	.5612	.6445	.7279	8112	.8945	.9779
1	.5625	.6458	.7292	.8125	.8958	.9792
	5638	.6471	.7305	.8138	.8971	.9805
	.5651	.6484	.7318	.8151	.8984	.9818
	.5664	.6497	.7331	.8164	.8997	.9831
	.5677	.6510	.7344	.8177	.9010	.9844
	.5690	.6523	.7357	.8190	.9023	.9857
	.5703	.6536	.7370	.8203	.9036	.9870
	5716	.6549	.7383	.8216	.9049	.9883
	.5729	.6562	.7396	.8229	.9062	.9896
	.5742	.6576	.7409	.8242	.9076	.9909
	.5755	.6589	.7422	.8255	.9089	.9922
	.5768	.6602	.7435	.8268	.9102	.9935
	.5781	.6615	.7448	.8281	9115	.9948
	5794	.6628	.7461	.8294	.9128	.9961
	.5807	.6641	.7474	.8307	.9141	.9974
	.5820	.6654	.7487	.8320	.9154	.9987
						1.0000

DECIMALS OF AN INCH FOR EACH 1/64th.

32ds.	64ths.	Decimal.	Fraction	32ds.	64ths.	Decimal.	Fraction
1	1	.015625			33	.515625	
	2	.03125		17	34	.53125	
	3	.046875			35	.546875	
2	4	.0625	1-16	18	36	.5625	9-16
	5	.078125			37	.578125	
3	6	.09375		19	38	.59375	
	7	.109375			39	.609375	
4	8	.125	1-8	20	40	.625	5-8
	9	.140625			41	.640625	
5	10	.15625		21	42	.65625	
	11	.171875			43	.671875	
6	12	.1875	3-16	22	44	.6875	11-16
	13	.203125			45	.703125	
7	14	.21875		23	46	.71875	
	15	.234375			47	.734375	
8	16	.25	1-4	24	48	.75	3-4
	17	.265625			49	.765625	
9	18	.28125		25	50	.78125	
	19	.296875			51	.796875	
10	20	.3125	5-16	26	52	.8125	13-16
	21	.328125			53	.828125	
11	22	.34375		27	54	.84375	
	23	.359375			55	.859375	
12	24	.375	3-8	28	56	.875	7-8
	25	.390625			57	.890625	
13	26	.40625		29	58	.90625	
	27	.421875			59	.921875	
14	28	.4375	7-16	30	60	.9375	15-16
	29	.453125			61	.953125	
15	30	.46875		31	62	.96875	
	31	.484375			63	.984375	
16	32	.5	1-2	32	64	1.	1

TABLES FOR CALCULATING THE HORSE POWER OF WATER.

MINERS' INCH TABLE.

The following table gives the Horse-Power of one miner's inch of water under heads from one up to eleven hundred feet. This inch equals 1½ cubic feet per minute.

CUBIC FEET TABLE.

The following table gives the Horse-Power of one cubic foot of water per minute under heads from one up to eleven hundred feet.

Heads in Feet.	Horse Power.	Heads in Feet.	Horse Power.	Heads in Feet.	Horse Power.	Heads in Feet.	Horse Power.
1	.0024147	320	.772704	1	.0016098	320	.515136
20	.0482294	330	.796851	20	.032196	330	.531234
30	.072441	340	.820998	30	.048294	340	.547332
40	.096588	350	.845145	40	.064392	350	.563430
50	.120735	360	.869292	50	.080490	360	.579528
60	.144882	370	.896439	60	.096588	370	.595626
70	.169029	380	.917586	70	.112686	380	.611724
80	.193176	390	.941733	80	.128784	390	.627822
90	.217323	400	.965880	90	.144892	400	.643920
100	.241470	410	.990027	100	.160980	410	.660018
110	.265617	420	1.014174	110	.177078	420	.676116
120	.289764	430	1.038321	120	.193176	430	.692214
130	.313911	440	1.062468	130	.209274	440	.708312
140	.338058	450	1.086615	140	.225372	450	.724410
150	.362205	460	1.110762	150	.241470	460	.740508
160	.386352	470	1.134909	160	.257568	470	.756606
170	.410499	480	1.159056	170	.273666	480	.772704
180	.434646	490	1.183206	180	.289764	490	.788802
190	.458793	500	1.207350	190	.305862	500	.804900
200	.482940	520	1.255644	200	.321960	520	.837096
210	.507087	540	1.303938	210	.338058	540	.869292
220	.531234	560	1.352232	220	.354156	560	.901488
230	.555381	580	1.400526	230	.370254	580	.933684
240	.579528	600	1.448820	240	.386352	600	.965880
250	.603675	650	1.569555	250	.402450	650	1.046370
260	.627822	700	1.690290	260	.418548	700	1.126860
270	.651969	750	1.811025	270	.434646	750	1.207350
280	.676116	800	1.931760	280	.450744	800	1.287840
290	.700263	900	2.173230	290	.466842	900	1.448820
300	.724410	1000	2.414700	300	.482940	1000	1.609800
310	.748557	1100	2.656170	310	.499038	1100	1.770780

WHEN THE EXACT HEAD IS FOUND IN ABOVE TABLE.

EXAMPLE.—Have 100 foot head and 50 inches of water. How many Horse-Power?

By reference to above table the Horse Power of 1 inch under 100 ft. head is .241470. This amount multiplied by the number of inches, 50, will give 12.07 Horse Power.

WHEN EXACT HEAD IS NOT FOUND IN TABLE.

Take the Horse Power of 1 inch under 1 ft. head and multiply by the number of inches, and then by number of feet head. The product will be the required Horse Power.

The above formula will answer for the cubic feet table, by substituting the the equivalents therein for those of miner's inches.

NOTE.—The above tables are based upon an efficiency of 85%.

LOSS OF HEAD IN PIPE BY FRICTION.

The following tables show the loss of head by friction in each 100 feet in length of different diameters of pipe when discharging the following quantities of water per minute:

INSIDE DIAMETER OF PIPE IN INCHES.

Velo in ft. per sec.	1 Loss of head in feet.	1 Cubic feet per min.	2 Loss of head in feet.	2 Cubic feet per min.	3 Loss of head in feet	3 Cubic feet per min.	4 Loss of head in feet.	4 Cubic feet per min.	5 Loss of head in feet.	5 Cubic feet per min.	6 Loss of head in feet.	6 Cubic feet per min.
2.0	2.37	.65	1.185	2.62	.791	5.89	.593	10.4	.474	16.3	.395	23.5
2.2	2.80	.73	1.404	2.88	.936	6.48	.702	11.5	.561	18.	.468	25.9
2.4	3.27	.79	1.639	3.14	1.093	7.07	.819	12.5	.650	19.6	.547	28.2
2.6	3.78	.86	1.891	3.40	1.26	7.65	.945	13.6	.757	21.3	.631	30.6
2.8	4.32	.92	2.16	3.66	1.44	8.24	1.08	14.6	.864	22.9	.720	32.9
3.0	4.89	.99	2.44	3.92	1.62	8.83	1.22	15.7	.978	24.5	.815	35.3
3.2	5.47	1.06	2.73	4.18	1.82	9.42	1.37	16.7	1.098	26.2	.915	37.7
3.4	6.09	1.12	3.05	4.45	2.04	10.00	1.52	17.8	1.22	27.8	1.021	40.
3.6	6.76	1.19	3.38	4.71	2.26	10.60	1.69	18.8	1.35	29.4	1.131	42.4
3.8	7.48	1.26	3.74	4.97	2.49	11.20	1.87	19.9	1.49	31.	1.25	44.7
4.0	8.20	1.32	4.10	5.23	2.73	11.80	2.05	20.9	1.64	32.7	1.37	47.1
4.2	8.97	1.39	4.49	5.49	2.98	12.30	2.24	22.0	1.79	34.3	1.49	49.5
4.4	9.77	1.45	4.89	5.76	3.25	12.90	2.43	23.0	1.95	36.0	1.62	51.8
4.6	10.60	1.52	5.30	6.02	3.53	13.50	2.64	24.0	2.11	37.6	1.76	54.1
4.8	11.45	1.58	5.72	6.28	3.81	14.10	2.85	25.1	2.27	39.2	1.90	56.5
5.0	12.33	1.65	6.17	6.54	4.11	14.70	3.08	26.2	2.46	40.9	2.05	58.9
5.2	13.24	1.72	6.62	6.80	4.41	15.30	3.31	27.2	2.65	42.5	2.21	61.2
5.4	14.20	1.78	7.10	7.06	4.73	15.90	3.55	28.2	2.84	44.2	2.37	63.6
5.6	15.16	1.85	7.58	7.32	5.06	16.50	3.79	29.3	3.03	45.8	2.53	65.9
5.8	16.17	1.91	8.09	7.58	5.40	17.10	4.04	30.3	3.24	47.4	2.70	68.3
6.0	17.23	1.98	8.61	7.85	5.74	17.70	4.31	31.4	3.45	49.1	2.87	70.7
7.0	22.89	2.31	11.45	9.16	7.62	20.6	5.72	36.6	4.57	57.2	3.81	82.4

INSIDE DIAMETER OF PIPE IN INCHES.

Velo in ft. per sec.	7 Loss of head in feet.	7 Cubic feet per min.	8 Loss of head in feet.	8 Cubic feet per min.	9 Loss of head in feet.	9 Cubic feet per min.	10 Loss of head in feet.	10 Cubic feet per min.	11 Loss of head in feet.	11 Cubic feet per min.	12 Loss of head in feet.	12 Cubic feet per min.
2.0	.338	32.0	.296	41.9	.264	53.	.237	65.4	.216	70.2	.198	94.2
2.2	.401	35.3	.351	46.1	.312	58.3	.281	72.	.255	87.1	.234	103.
2.4	.468	38.5	.410	50.2	.365	63.6	.327	78.5	.297	95.0	.273	113.
2.6	.540	41.7	.473	54.4	.420	68.9	.378	85.1	.344	103.	.315	122.
2.8	.617	44.9	.540	58.6	.480	74.2	.432	91.6	.392	111.	.360	132.
3.0	.698	48.1	.611	62.8	.544	79.5	.488	98.2	.444	119.	.407	141.
3.2	.785	51.3	.686	67.	.609	84.8	.549	105.	.499	127.	.457	151.
3.4	.875	54.5	.765	71.2	.680	90.1	.612	111.	.557	134.	.510	160.
3.6	.969	57.7	.848	75.4	.755	95.4	.679	118.	.617	142.	.566	169.
3.8	1.070	60.9	.936	79.6	.831	101.	.749	124.	.680	150.	.624	179.
4.0	1.175	64.1	1.027	83.7	.913	106.	.822	131.	.747	158.	.685	188.
4.2	1.28	67.3	1.122	87.9	.998	111.	.897	137.	.816	166.	.749	198.
4.4	1.39	70.5	1.22	92.1	1.086	116.	.977	144.	.888	174.	.815	207.
4.6	1.51	73.7	1.32	96.3	1.177	122.	1.059	150.	.963	182.	.883	217.
4.8	1.63	76.9	1.43	100.0	1.27	127.	1.145	157.	1.040	190.	.954	226.
5.0	1.76	80.2	1.54	105.	1.37	132.	1.23	163.	1.122	198.	1.028	235.
5.2	1.89	83.3	1.65	109.	1.47	138.	1.32	170.	1.20	206.	1.104	245.
5.4	2.03	86.6	1.77	113.	1.57	143.	1.41	177.	1.28	214.	1.183	254.
5.6	2.17	89.8	1.80	117.	1.68	148.	1.51	183.	1.37	222.	1.26	264.
5.8	2.31	93.0	2.01	121.	1.80	154.	1.61	190.	1.46	229.	1.34	273.
6.0	2.46	96.2	2.15	125.	1.92	159.	1.71	196.	1.56	237.	1.43	283.
7.0	3.26	112.0	2.85	146.	2.52	185.	2.28	229.	2.07	277.	1.91	330.

LOSS OF HEAD IN PIPE BY FRICTION.

The following tables show the loss of head by friction in each 100 feet in length of different diameters of pipe when discharging the following quantities of water per minute:

INSIDE DIAMETER OF PIPE IN INCHES.

Velo. in ft. per sec.	13		14		15		16		18		20	
	Loss of head in feet.	Cubic feet per min.	Loss of head in feet.	Cubic feet per min.	Loss of head in feet.	Cubic feet per min.	Loss of head in feet.	Cubic feet per min.	Loss of head in feet.	Cubic feet per min.	Loss of head in feet.	Cubic feet per min.
2.0	.183	110.	.169	128.	.158	147.	.147	167.	.132	212.	.119	262.
2.2	.216	121.	.200	141.	.187	162.	.175	184.	.156	233.	.140	288.
2.4	.252	133.	.234	154.	.218	176.	.205	201.	.182	254.	.164	314.
2.6	.290	144.	.270	167.	.252	191.	.236	218.	.210	275.	.189	340.
2.8	.332	156.	.308	179.	.288	206.	.270	234.	.240	297.	.216	366.
3.0	.375	166.	.349	192.	.325	221.	.306	251.	.271	318.	.245	393.
3.2	.422	177.	.392	205.	.366	235.	.343	268.	.305	339.	.275	419.
3.4	.471	188.	.438	218.	.408	250.	.383	284.	.339	360.	.306	445.
3.6	.522	199.	.485	231.	.452	265.	.425	301.	.377	382.	.339	471.
3.8	.576	210.	.535	243.	.499	280.	.468	318.	.416	403.	.374	497.
4.0	.632	221.	.597	256.	.548	294.	.513	335.	.456	424.	.410	523.
4.2	.691	232.	.641	269.	.598	309.	.561	352.	.499	445.	.449	550.
4.4	.751	243.	.698	282.	.651	324.	.611	368.	.542	466.	.488	576.
4.6	.815	254.	.757	295.	.707	339.	.662	385.	.588	488.	.529	602.
4.8	.881	265.	.818	308.	.763	353.	.715	402.	.636	509.	.572	628.
5.0	.949	276.	.881	321.	.822	368.	.770	419.	.685	530.	.617	654.
5.2	1.020	287.	.947	333.	.883	383.	.828	435.	.736	551.	.662	680.
5.4	1.092	298.	1.014	346.	.947	397.	.886	452.	.788	572.	.710	707.
5.6	1.167	309.	1.083	359.	1.011	412.	.949	469.	.843	594.	.758	733.
5.8	1.245	321.	1.155	372.	1.078	427.	1.011	486.	.899	615.	.809	759.
6.0	1.325	332.	1.229	385.	1.148	442.	1.076	502.	.957	636.	.861	785.
7.0	1.75	387.	1.63	449.	1.52	515.	1.43	586.	1.27	742.	1.143	916.

INSIDE DIAMETER OF PIPE IN INCHES.

Velo. in ft. per sec.	23		24		26		28		30		36	
	Loss of head in feet.	Cubic feet per min.	Loss of head in feet.	Cubic feet per min.	Loss of head in feet.	Cubic feet per min.	Loss of head in feet.	Cubic feet per min.	Loss of head in feet.	Cubic feet per min.	Loss of head in feet.	Cubic feet per min.
2.0	.108	316.	.098	377.	.091	442.	.084	513.	.079	589.	.066	848.
2.2	.127	348.	.116	414.	.108	486.	.099	564.	.093	648.	.078	933.
2.4	.149	380.	.136	452.	.126	531.	.116	616.	.109	707.	.091	1018.
2.6	.171	412.	.157	490.	.145	575.	.134	667.	.126	766.	.104	1100.
2.8	.195	443.	.180	528.	.165	619.	.153	718.	.144	824.	.119	1188.
3.0	.222	475.	.204	565.	.188	663.	.174	770.	.163	883.	.135	1273.
3.2	.249	507.	.229	603.	.211	708.	.195	821.	.182	942.	.152	1357.
3.4	.278	538.	.255	641.	.235	752.	.218	872.	.204	1001.	.169	1442.
3.6	.308	570.	.283	678.	.261	796.	.242	923.	.226	1060.	.188	1527.
3.8	.340	601.	.312	716.	.288	840.	.267	974.	.249	1119.	.207	1612.
4.0	.373	633.	.342	754.	.315	885.	.293	1026.	.273	1178.	.228	1697.
4.2	.408	665.	.374	791.	.345	929.	.320	1077.	.299	1237.	.249	1782.
4.4	.444	697.	.407	829.	.375	973.	.348	1129.	.325	1296.	.271	1866.
4.6	.482	728.	.441	867.	.407	1017.	.378	1180.	.353	1355.	.294	1951.
4.8	.521	760.	.476	905.	.440	1062.	.409	1231.	.381	1414.	.318	2036.
5.0	.561	792.	.513	942.	.474	1106.	.440	1283.	.411	1472.	.312	2121.
5.2	.602	823.	.552	980.	.510	1150.	.473	1334.	.441	1531.	.368	2200.
5.4	.645	855.	.591	1018.	.546	1194.	.507	1885.	.473	1590.	.394	2291.
5.6	.690	887.	.632	1055.	.583	1239.	.542	1437.	.506	1649.	.421	2376.
5.8	.735	918.	.674	1093.	.622	1283.	.578	1488.	.540	1708.	.450	2460.
6.0	.782	950.	.717	1131.	.662	1327.	.615	1539.	.574	1767.	.479	2545.
7.0	1.040	1109.	.953	1319.	.879	1548.	.817	1796.	.762	2061.	.636	2968.

TABLE OF SHEET IRON HYDRAULIC PIPE.

Diameter of pipe in inches.	Area of pipe in inches.	Thickn's of iron by wire gauge.	Head in feet the pipe will safely stand.	Cub. ft. of water pipe will convey per min. at vel. 3 ft. per second.	Weight per lineal foot in lbs.
3	7	18	400	9	2
4	12	18	350	16	2½
4	12	16	525	16	3
5	20	18	325	25	3½
5	20	16	500	25	4½
5	20	14	675	25	5
6	28	18	206	36	4½
6	28	16	487	36	5¾
6	28	14	743	36	7½
7	38	18	254	50	5¼
7	38	16	419	50	6¾
7	38	14	640	50	8½
8	50	16	367	63	7½
8	50	14	560	63	9¾
8	50	12	854	63	13
9	63	16	327	80	8¼
9	63	14	499	80	10¾
9	63	12	761	80	14½
10	78	16	295	100	9½
10	78	14	450	100	11¾
10	78	12	687	100	15¾
10	78	11	754	100	17¾
10	78	10	900	100	19½
11	95	16	269	120	9¾
11	95	14	412	120	13
11	95	12	626	120	17½
11	95	11	687	120	18¾
11	95	10	820	120	21
12	113	16	246	142	11½
12	113	14	377	142	14
12	113	12	574	142	18¾
12	113	11	630	142	19¾
12	118	10	758	142	22¾
13	132	16	228	170	12
13	132	14	348	170	15
13	132	12	530	170	20
13	132	11	583	170	22
13	132	10	696	170	24½
14	153	16	211	200	13
14	153	14	324	200	16
14	153	12	494	200	21½
14	153	11	543	200	23½
14	153	10	648	200	26
15	176	16	197	225	13¾
15	176	14	302	225	17
15	176	12	460	225	23
15	176	11	507	225	24½
15	176	10	606	225	28
16	201	16	185	255	14½
16	201	14	283	255	17½
16	201	12	432	255	24½
16	201	11	474	255	26½
16	201	10	567	255	29½
18	254	16	165	320	10½
18	254	14	252	320	20½
18	254	12	385	320	27¼
18	254	11	424	320	30
18	254	10	505	320	34
20	314	16	148	400	18
20	314	14	227	400	22½
20	314	12	346	400	30
20	314	11	380	400	32½
20	314	10	456	400	36½
22	380	16	135	480	20
22	380	14	206	480	24¾
22	380	12	316	480	32¾
22	380	11	347	480	35½
22	380	10	415	480	40
24	452	14	188	570	27½
24	452	12	290	570	35½
24	452	11	318	570	39
24	452	10	379	570	43½
24	452	8	466	570	53
26	530	14	175	670	29½
26	530	12	267	670	38½
26	530	11	294	670	42
26	530	10	352	670	47
26	530	8	432	670	57½
28	615	14	102	775	31½
28	615	12	247	775	41½
28	615	11	273	775	45
28	615	10	327	775	50½
28	615	8	400	775	61½
30	706	12	231	890	44
30	706	11	254	890	48
30	706	10	304	890	54
30	706	8	375	890	65
30	706	7	425	890	74
33	1017	11	141	1300	58
33	1017	10	155	1300	67
36	1017	8	192	1300	78
36	1017	7	210	1300	88
40	1256	10	141	1600	71
40	1256	8	174	1600	86
40	1256	7	189	1600	97
40	1256	6	213	1600	108
40	1256	4	250	1600	126
42	1385	10	135	1760	74½
42	1385	8	165	1760	91
42	1385	7	180	1760	102
42	1385	6	210	1760	114
42	1385	4	240	1760	133
42	1385	½	270	1760	137
42	1385	3	300	1760	145
42	1385	1/16	321	1760	177
42	1385	¼	363	1760	216

Table of Water Wheel Velocity.

Head in Feet	Spouting Velocity Feet per Minute	Head in Feet	Spouting Velocity Feet per Minute	Head in Feet	Spouting Velocity Feet per Minute	Head in Feet	Spouting Velocity Feet per Minute
20	2152	170	6274	320	8608	470	10432
30	2636	180	6456	330	8741	480	10543
40	3043	190	6633	340	8873	490	10652
50	3403	200	6805	350	9002	500	10760
60	3727	210	6973	360	9130	520	10973
70	4026	220	7137	370	9256	540	11182
80	4304	230	7298	380	9380	560	11387
90	4565	240	7455	390	9503	580	11589
100	4812	250	7608	400	9624	600	11787
110	5047	260	7759	410	9744	650	12268
120	5271	270	7907	420	9862	700	12731
130	5487	280	8052	430	9978	750	13178
140	5694	290	8195	440	10094	800	13610
150	5893	300	8335	450	10208	900	14436
160	6087	310	8472	460	10321	1000	15217

——— Calling: S the Spouting Velocity in Feet per Minute,
D the Diameter of the Wheel in Feet,
N the Number of Revolutions of Wheel per Minute,
——— the Diameter is found from: $D = S \div (6.28 \times N)$
——— the Number of Revolutions per Minute is found from: $N = S \div (6.28 \times D)$

The following is a very useful table and should be employed in Compressed Air distribution. The efficiency of many plants would be increased if the piping followed these proportions:

Equation of Pipes.—It is frequently desired to know what number of pipes of a given size are equal in carrying capacity to one pipe of a larger size. At the same velocity of flow the volume delivered by two pipes of different sizes is proportional to the squares of their diameters; thus, one 4-inch pipe will deliver the same volume as four 2-inch pipes. With the same head, however, the velocity is less in the smaller pipe, and the volume delivered varies about as the square root of the fifth power (i.e., as the 2.5 power). The following table has been calculated on this basis. The figures opposite the intersection of any two sizes is the number of the smaller-sized pipes required to equal one of the larger. Thus, one 4-inch pipe is equal to 5.7 2-inch pipes.

Diam. in.	1	2	3	4	5	6	7	8	9	10	12	14	16	18	20	24
2	5.7	1														
3	15.6	2.8	1													
4	32	5.7	2.1	1												
5	55.9	9.9	3.6	1.7	1											
6	88.2	15.6	5.7	2.8	1.6	1										
7	130	22.9	8.3	4.1	2.3	1.5	1									
8	181	32	11.7	5.7	3.2	2.1	1.4	1								
9	248	43	15.6	7.6	4.3	2.8	1.9	1.3	1							
10	316	55.9	20.3	9.9	5.7	3.6	2.4	1.7	1.3	1						
11	401	70.9	25.7	12.5	7.2	4.6	3.1	2.2	1.7	1.3						
12	499	88.2	32	15.6	8.9	5.7	3.8	2.8	2.1	1.6	1					
13	609	108	39.1	19	10.9	7.1	4.7	3.4	2.5	1.9	1.2					
14	733	130	47	22.9	13.1	8.3	5.7	4.1	3.0	2.3	1.5	1				
15	787	154	55.9	27.2	15.6	9.9	6.7	4.8	3.6	2.8	1.7	1.2				
16	..	181	65.7	32	18.3	11.7	7.9	5.7	4.2	3.2	2.1	1.4	1			
17	...	211	76.4	37.2	21.3	13.5	9.2	6.6	4.9	3.8	2.4	1.6	1.2			
18	...	243	88.2	43	24.6	15.6	10.6	7.6	5.7	4.3	2.8	1.9	1.3	1		
19	...	278	101	49.1	28.1	17.8	12.1	8.7	6.5	5	3.2	2.1	1.5	1.1		
20	...	316	115	55.9	32	20.3	13.8	9.9	7.4	5.7	3.6	2.4	1.7	1.3	1	
22	...	401	146	70.9	40.6	25.7	17.5	12.5	9.3	7.2	4.6	3.1	2.2	1.7	1.3	
24	...	499	181	88.2	50.5	32	21.8	15.6	11.6	8.9	5.7	3.8	2.8	2.1	1.6	1
26	...	609	221	108	61.7	39.1	26.6	19.	14.2	10.9	7.1	4.7	3.4	2.5	1.9	1.2
28	...	733	266	130	74.2	47	32	22.9	17.1	13.1	8.3	5.7	4.1	3	2.3	1.5
30	...	787	316	154	88.2	55.9	38	27.2	20.3	15.6	9.9	6.7	4.8	3.6	2.8	1.7
36			499	243	130	88.2	60	43	32	24.6	15.6	10.6	7.6	5.7	4.3	2.8
42			733	357	205	130	88.2	63.2	47	36.2	19	15.6	11.2	8.3	6.4	4.1
48			...	499	286	181	123	88.2	62.7	50.5	32	21.8	15.6	11.6	8.9	5.7
54				670	383	243	165	118	88.2	67.8	43	29.2	20.9	15.6	12	7.6
60				787	499	316	215	154	115	88.2	55.9	38	27.2	20.3	15.6	9.9

Used in the calculation of problems in Isothermal Compression and Expausion of Compressed Air.

HYPERBOLIC LOGARITHMS.

No.	Log.	No.	Log.	No.	Log.	No.	Log.	No.	Log.
1.01	.0099	1.45	.3716	1.89	.6366	2.33	.8458	2.77	1.0188
1.02	.0198	1.46	.3784	1.90	.6419	2.34	.8502	2.78	1.0225
1.03	.0296	1.47	.3853	1.91	.6471	2.35	.8544	2.79	1.0260
1.04	.0392	1.48	.3920	1.92	.6523	2.36	.8587	2.80	1.0296
1.05	.0488	1.49	.3988	1.93	.6575	2.37	.8629	2.81	1.0332
1.06	.0583	1.50	.4055	1.94	.6627	2.38	.8671	2.82	1.0367
1.07	.0677	1.51	.4121	1.95	.6678	2.39	.8713	2.83	1.0403
1.08	.0770	1.52	.4187	1.96	.6729	2.40	.8755	2.84	1.0438
1.09	.0862	1.53	.4253	1.97	.6780	2.41	.8796	2.85	1.0473
1.10	.0953	1.54	.4318	1.98	.6831	2.42	.8838	2.86	1.0508
1.11	.1044	1.55	.4383	1.99	.6881	2.43	.8879	2.87	1.0543
1.12	.1133	1.56	.4447	2.00	.6931	2.44	.8920	2.88	1.0578
1.13	.1222	1.57	.4511	2.01	.6981	2.45	.8961	2.89	1.0613
1.14	.1310	1.58	.4574	2.02	.7031	2.46	.9002	2.90	1.0647
1.15	.1398	1.59	.4637	2.03	.7080	2.47	.9042	2.91	1.0682
1.16	.1484	1.60	.4700	2.04	.7129	2.48	.9083	2.92	1.0716
1.17	.1570	1.61	.4762	2.05	.7178	2.49	.9123	2.93	1.0750
1.18	.1655	1.62	.4824	2.06	.7227	2.50	.9163	2.94	1.0784
1.19	.1740	1.63	.4886	2.07	.7275	2.51	.9203	2.95	1.0818
1.20	.1823	1.64	.4947	2.08	.7324	2.52	.9243	2.96	1.0852
1.21	.1906	1.65	.5008	2.09	.7372	2.53	.9282	2.97	1.0886
1.22	.1988	1.66	.5068	2.10	.7419	2.54	.9322	2.98	1.0919
1.23	.2070	1.67	.5128	2.11	.7467	2.55	.9361	2.99	1.0953
1.24	.2151	1.68	.5188	2.12	.7514	2.56	.9400	3.00	1.0986
1.25	.2231	1.69	.5247	2.13	.7561	2.57	.9439	3.01	1.1019
1.26	.2311	1.70	.5306	2.14	.7608	2.58	.9478	3.02	1.1053
1.27	.2390	1.71	.5365	2.15	.7655	2.59	.9517	3.03	1.1086
1.28	.2469	1.72	.5423	2.16	.7701	2.60	.9555	3.04	1.1119
1.29	.2546	1.73	.5481	2.17	.7747	2.61	.9594	3.05	1.1151
1.30	.2624	1.74	.5539	2.18	.7793	2.62	.9632	3.06	1.1184
1.31	.2700	1.75	.5596	2.19	.7839	2.63	.9670	3.07	1.1217
1.32	.2776	1.76	.5653	2.20	.7885	2.64	.9708	3.08	1.1249
1.33	.2852	1.77	.5710	2.21	.7930	2.65	.9746	3.09	1.1282
1.34	.2927	1.78	.5766	2.22	.7975	2.66	.9783	3.10	1.1314
1.35	.3001	1.79	.5822	2.23	.8020	2.67	.9821	3.11	1.1346
1.36	.3075	1.80	.5878	2.24	.8065	2.68	.9858	3.12	1.1378
1.37	.3148	1.81	.5933	2.25	.8109	2.69	.9895	3.13	1.1410
1.38	.3221	1.82	.5988	2.26	.8154	2.70	.9933	3.14	1.1442
1.39	.3293	1.83	.6043	2.27	.8198	2.71	.9969	3.15	1.1474
1.40	.3365	1.84	.6098	2.28	.8242	2.72	1.0006	3.16	1.1506
1.41	.3436	1.85	.6152	2.29	.8286	2.73	1.0043	3.17	1.1537
1.42	.3507	1.86	.6206	2.30	.8329	2.74	1.0080	3.18	1.1569
1.43	.3577	1.87	.6259	2.31	.8372	2.75	1.0116	3.19	1.1600
1.44	.3646	1.88	.6313	2.32	.8416	2.76	1.0152	3.20	1.1632

HYPERBOLIC LOGARITHMS.

No.	Log.	No.	Log.	No.	Log.	No.	Log.	No.	Log.
3.21	1.1663	3.87	1.3533	4.53	1.5107	5.19	1.6467	5.85	1.7664
3.22	1.1694	3.88	1.3558	4.54	1.5129	5.20	1.6487	5.86	1.7681
3.23	1.1725	3.89	1.3584	4.55	1.5151	5.21	1.6506	5.87	1.7699
3.24	1.1756	3.90	1.3610	4.56	1.5173	5.22	1.6525	5.88	1.7716
3.25	1.1787	3.91	1.3635	4.57	1.5195	5.23	1.6544	5.89	1.7733
3.26	1.1817	3.92	1.3661	4.58	1.5217	5.24	1.6563	5.90	1.7750
3.27	1.1848	3.93	1.3686	4.59	1.5239	5.25	1.6582	5.91	1.7766
3.28	1.1878	3.94	1.3712	4.60	1.5261	5.26	1.6601	5.92	1.7783
3.29	1.1909	3.95	1.3737	4.61	1.5282	5.27	1.6620	5.93	1.7800
3.30	1.1939	3.96	1.3762	4.62	1.5304	5.28	1.6639	5.94	1.7817
3.31	1.1969	3.97	1.3788	4.63	1.5326	5.29	1.6658	5.95	1.7834
3.32	1.1999	3.98	1.3813	4.64	1.5347	5.30	1.6677	5.96	1.7851
3.33	1.2030	3.99	1.3838	4.65	1.5369	5.31	1.6696	5.97	1.7867
3.34	1.2060	4.00	1.3863	4.66	1.5390	5.32	1.6715	5.98	1.7884
3.35	1.2090	4.01	1.3888	4.67	1.5412	5.33	1.6734	5.99	1.7901
3.36	1.2119	4.02	1.3913	4.68	1.5433	5.34	1.6752	6.00	1.7918
3.37	1.2149	4.03	1.3938	4.69	1.5454	5.35	1.6771	6.01	1.7934
3.38	1.2179	4.04	1.3962	4.70	1.5476	5.36	1.6790	6.02	1.7951
3.39	1.2208	4.05	1.3987	4.71	1.5497	5.37	1.6808	6.03	1.7967
3.40	1.2238	4.06	1.4012	4.72	1.5518	5.38	1.6827	6.04	1.7984
3.41	1.2267	4.07	1.4036	4.73	1.5539	5.39	1.6845	6.05	1.8001
3.42	1.2296	4.08	1.4061	4.74	1.5560	5.40	1.6864	6.06	1.8017
3.43	1.2326	4.09	1.4085	4.75	1.5581	5.41	1.6882	6.07	1.8034
3.44	1.2355	4.10	1.4110	4.76	1.5602	5.42	1.6901	6.08	1.8050
3.45	1.2384	4.11	1.4134	4.77	1.5623	5.43	1.6919	6.09	1.8066
3.46	1.2413	4.12	1.4159	4.78	1.5644	5.44	1.6938	6.10	1.8083
3.47	1.2442	4.13	1.4183	4.79	1.5665	5.45	1.6956	6.11	1.8099
3.48	1.2470	4.14	1.4207	4.80	1.5686	5.46	1.6974	6.12	1.8116
3.49	1.2499	4.15	1.4231	4.81	1.5707	5.47	1.6993	6.13	1.8132
3.50	1.2528	4.16	1.4255	4.82	1.5728	5.48	1.7011	6.14	1.8148
3.51	1.2556	4.17	1.4279	4.83	1.5748	5.49	1.7029	6.15	1.8165
3.52	1.2585	4.18	1.4303	4.84	1.5769	5.50	1.7047	6.16	1.8181
3.53	1.2613	4.19	1.4327	4.85	1.5790	5.51	1.7066	6.17	1.8197
3.54	1.2641	4.20	1.4351	4.86	1.5810	5.52	1.7084	6.18	1.8213
3.55	1.2669	4.21	1.4375	4.87	1.5831	5.53	1.7102	6.19	1.8229
3.56	1.2698	4.22	1.4398	4.88	1.5851	5.54	1.7120	6.20	1.8245
3.57	1.2726	4.23	1.4422	4.89	1.5872	5.55	1.7138	6.21	1.8262
3.58	1.2754	4.24	1.4446	4.90	1.5892	5.56	1.7156	6.22	1.8278
3.59	1.2782	4.25	1.4469	4.91	1.5913	5.57	1.7174	6.23	1.8294
3.60	1.2809	4.26	1.4493	4.92	1.5933	5.58	1.7192	6.24	1.8310
3.61	1.2837	4.27	1.4516	4.93	1.5953	5.59	1.7210	6.25	1.8326
3.62	1.2865	4.28	1.4540	4.94	1.5974	5.60	1.7228	6.26	1.8342
3.63	1.2892	4.29	1.4563	4.95	1.5994	5.61	1.7246	6.27	1.8358
3.64	1.2920	4.30	1.4586	4.96	1.6014	5.62	1.7263	6.28	1.8374
3.65	1.2947	4.31	1.4609	4.97	1.6034	5.63	1.7281	6.29	1.8390
3.66	1.2975	4.32	1.4633	4.98	1.6054	5.64	1.7299	6.30	1.8405
3.67	1.3002	4.33	1.4656	4.99	1.6074	5.65	1.7317	6.31	1.8421
3.68	1.3029	4.34	1.4679	5.00	1.6094	5.66	1.7334	6.32	1.8437
3.69	1.3056	4.35	1.4702	5.01	1.6114	5.67	1.7352	6.33	1.8453
3.70	1.3083	4.36	1.4725	5.02	1.6134	5.68	1.7370	6.34	1.8469
3.71	1.3110	4.37	1.4748	5.03	1.6154	5.69	1.7387	6.35	1.8485
3.72	1.3137	4.38	1.4770	5.04	1.6174	5.70	1.7405	6.36	1.8500
3.73	1.3164	4.39	1.4793	5.05	1.6194	5.71	1.7422	6.37	1.8516
3.74	1.3191	4.40	1.4816	5.06	1.6214	5.72	1.7440	6.38	1.8532
3.75	1.3218	4.41	1.4839	5.07	1.6233	5.73	1.7457	6.39	1.8547
3.76	1.3244	4.42	1.4861	5.08	1.6253	5.74	1.7475	6.40	1.8563
3.77	1.3271	4.43	1.4884	5.09	1.6273	5.75	1.7492	6.41	1.8579
3.78	1.3297	4.44	1.4907	5.10	1.6292	5.76	1.7509	6.42	1.8594
3.79	1.3324	4.45	1.4929	5.11	1.6312	5.77	1.7527	6.43	1.8610
3.80	1.3350	4.46	1.4951	5.12	1.6332	5.78	1.7544	6.44	1.8625
3.81	1.3376	4.47	1.4974	5.13	1.6351	5.79	1.7561	6.45	1.8641
3.82	1.3403	4.48	1.4996	5.14	1.6371	5.80	1.7579	6.46	1.8656
3.83	1.3429	4.49	1.5019	5.15	1.6390	5.81	1.7596	6.47	1.8672
3.84	1.3455	4.50	1.5041	5.16	1.6409	5.82	1.7613	6.48	1.8687
3.85	1.3481	4.51	1.5063	5.17	1.6429	5.83	1.7630	6.49	1.8703
3.86	1.3507	4.52	1.5085	5.18	1.6448	5.84	1.7647	6.50	1.8718

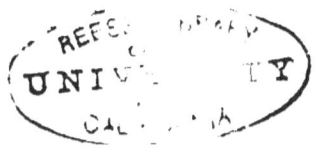

Volume, Density, and Pressure of Air at Various Temperatures. (D. K. Clark.)

Fahr.	Volume at Atmos. Pressure.		Density, lbs. per Cubic Foot at Atmos. Pressure.	Pressure at Constant Volume.	
	Cubic Feet in 1 lb.	Comparative Vol.		Lbs. per Sq. In.	Comparative Pres.
0	11.583	.881	.086331	12.96	.881
32	12.387	.943	.080728	13.86	.943
40	12.586	.958	.079439	14.08	.958
50	12.840	.977	.077884	14.36	.977
62	13.141	1.000	.076097	14.70	1.000
70	13.342	1.015	.074950	14.92	1.015
80	13.593	1.034	.073565	15.21	1.034
90	13.845	1.054	.072230	15.49	1.054
100	14.096	1.073	.070942	15.77	1.073
110	14.344	1.092	.069721	16.05	1.092
120	14.592	1.111	.068500	16.33	1.111
130	14.846	1.130	.067361	16.61	1.130
140	15.100	1.149	.066221	16.89	1.149
150	15.351	1.168	.065155	17.19	1.168
160	15.603	1.187	.064088	17.50	1.187
170	15.854	1.206	.063089	17.76	1.206
180	16.106	1.226	.062090	18.02	1.226
200	16.606	1.264	.060210	18.58	1.264
210	16.860	1.283	.059313	18.86	1.283
212	16.910	1.287	.059135	18.92	1.287

Volumes, Mean Pressures per Stroke, Temperatures, etc., in the Operation of Air-compression from 1 Atmosphere and 60° Fahr. (F. Richards, *Am. Mach.*, March 30, 1893.)

Gauge-pressure.	Atmospheres.	Volume with Air at Constant Temp.	Volume with Air not cooled.	Mean Pressure per Stroke; Air Constant Temp.	Mean Pressure per Stroke; Air not cooled.	Temp. of Air; not cooled.	Gauge-pressure.	Atmospheres.	Volume with Air at Constant Temp.	Volume with Air not cooled.	Mean Pressure per Stroke; Air Constant Temp.	Mean Pressure per Stroke; Air not cooled.	Temp. of Air; not cooled.
1	2	3	4	5	6	7	1	2	3	4	5	6	7
0	1	1	1	0	0	60°	80	6.442	.1552	.267	27.38	36.64	432
1	1.068	.9363	.95	.96	.975	71	85	6.782	.1474	.2566	28.16	37.94	447
2	1.136	.8803	.91	1.87	1.91	80.4	90	7.122	.1404	.248	28.89	39.18	459
3	1.204	.8305	.876	2.72	2 8	88.9	95	7.462	.134	.24	29.57	40.4	472
4	1.272	.7861	.84	3.53	3 67	96	100	7.802	.1281	.232	30.21	41.6	485
5	1.34	.7462	.81	4.3	4 5	106	105	8.142	.1228	.2254	30.81	42.78	496
10	1 68	.5952	.69	7.62	8.27	145	110	8 483	.1178	.2189	31.39	43.91	507
15	2.02	.495	.606	10.83	11.51	178	115	8 823	.1133	.2129	31.98	44.98	518
20	2.36	.4237	.543	12.62	14.4	207	120	9.163	.1091	.2073	32.54	46.04	529
25	2.7	.3703	.494	14.59	17.01	234	125	9.503	.1052	.2020	33.07	47.06	540
30	3.04	.3289	.4538	16.34	19.4	252	130	9.843	.1015	.1969	33.57	48.1	550
35	3.391	.2957	.42	17.92	21.6	261	135	10.183	.0981	.1922	34.05	49.1	560
40	3.721	.2687	.393	19.32	23 66	302	140	10.523	.095	.1878	34.57	50.02	570
45	4.061	.2162	.37	20.57	25.59	321	145	10.864	.0921	.1837	35.09	51.	580
50	4.401	.2272	.35	21.69	27.39	339	150	11.204	.0892	.1796	35.48	51.89	589
55	4.741	.2109	.331	22.76	29.11	357	160	11.88	.0841	.1722	36.29	53.65	607
60	5.081	.1968	.3144	23.78	30 75	375	170	12.56	0796	.1657	37.2	55.39	624
65	5.423	.1844	.301	24.75	32 32	389	180	13 24	0755	.1595	37.96	57.01	640
70	5.762	.1735	.288	25 67	33.83	405	190	13.92	.0718	.154	38.68	58.57	657
75	6.102	.1639	.276	26.55	35.27	420	200	14.6	.0685	.149	39.42	60.14	672

Mean and Terminal Pressures of Compressed Air used Expansively for Gauge-pressures from 60 to 100 lbs.

(Frank Richards, *Am. Mach.*, April 13, 1893.)

Initial Pressure. Point of Cut-off.	60.		70.		80.		90.		100.	
	Mean Air-pressure.	Terminal Air-pressure.	Mean Air-pressure.	Terminal Air-pressure.	Mean Air pressure.	Terminal Air-pressure.	Mean Air-pressure.	Terminal Air-pressure.	Mean Air-pressure	Terminal Air-pressure
.25	23.6	*10 65*	28.74	*12.07*	33.89	*13.49*	39 04	*14.91*	44.19	1.33
.30	28.9	*13.77*	34.75	6	40 61	2.44	46 46	4 27	53.82	6 11
⅓	32.13	.96	38.41	3.09	44.69	5 22	50.98	7 35	57 26	9 48
.35	33.66	2.33	40.75	4.38	46.64	6.66	53.13	8 95	59 62	11.23
⅜	35.85	3.85	42 63	6.36	49.41	7.88	56 2	11.89	62 98	13.59
.40	37.93	5.64	44.99	8.39	52.05	11.14	59 11	13.88	66 16	16.64
.45	41.75	10.71	49.31	12.61	56.9	15.86	64 45	19.11	72.02	22.36
.50	45.14	13.26	53.16	17	61.18	20 81	69.19	24 56	77.21	28.33
.60	50.75	21.53	59.51	26.4	68.28	31 27	77.05	36.14	85.82	41 01
⅝	51.92	23.69	60.84	28.85	69.76	34.01	78.69	39.16	87.61	44.32
⅔	53.67	27.94	62 83	33.03	71.99	38.68	81 14	44 33	90.32	49 97
.70	54.93	30.39	64 25	36 44	73.57	42.49	82 9	48.54	92.22	54.59
.75	56.52	35.01	66.05	41.68	75.59	48.35	85.12	55.02	94.66	61 69
.80	57.79	39.78	67.5	47.08	77.2	54.38	86.91	61.69	96.61	68.99
⅞	59.15	47.14	69.03	55.43	78.92	63.81	88.81	72.	98.7	80 28
.90	59.46	49.65	69.38	58.27	79 31	66 89	89.24	75.52	99.17	87 82

The pressures in the table are all gauge-pressures except those in *italics*, which are absolute pressures (above a vacuum).

R	$\frac{1}{R}$	$\frac{P_m}{P_1}$	R	$\frac{1}{R}$	$\frac{P_m}{P_1}$
20	.05	.1998	5	.2	.5218
18	.055	.2161	4.44	.225	.5608
16	.062	.2358	4.	.25	.5965
15	.066	.2472	3.63	.275	.6308
14	.071	.2599	3.33	.3	.6615
13.33	.075	.269	3	.333	.6993
13	.077	.2742	2.86	.35	.7171
12	.083	.2904	2.66	.375	.744
11	.091	.3089	2.50	.4	.7664
10	.1	.3303	2.22	.45	.8095
9	.111	.3552	2.	.5	.8465
8	.125	.3849	1.82	.55	.8786
7	.143	.421	1.66	.6	.9066
6.66	.15	.4347	1.60	.625	.9187
6	.166	.4653	1.54	.65	.9292
5.71	.175	.4807	1.48	.675	.9405

—— Table of Mean Absolute Pressures ——
—— for Various Degrees of Isothermal Expansion ——
—— In this Table ——

P_1 is the Absolute Pressure at which Steam enters the Cylinder.
P_m is the corresponding Mean Absolute Pressure.
R is the Rate of Expansion, i.e. the Ratio of the total Volume of Cylinder, including Clearance, to the Volume of Live Steam, including clearance.
$\frac{1}{R}$ indicates the point of Cut-off, i.e. the Ratio of the total Volume of Live Steam to the total Volume of Cylinder at the End of the Stroke.

Nominal Inside Diameter — Inches —	Actual Outside Diameter — Inches —	Nominal Weight per Foot — Lbs —	Number of Threads per Inch of Screw
2	2·$\frac{1}{4}$	2.22	14
2·$\frac{1}{4}$	2·$\frac{1}{2}$	2.82	14
2·$\frac{1}{2}$	2·$\frac{3}{4}$	3.13	14
2·$\frac{3}{4}$	3	3.45	14
3	3·$\frac{1}{4}$	4.10	14
3·$\frac{1}{4}$	3·$\frac{1}{2}$	4.45	14
3·$\frac{1}{2}$	3·$\frac{3}{4}$	4.78	14
3·$\frac{3}{4}$	4	5.56	14
4	4·$\frac{1}{4}$	6.	14
4·$\frac{1}{4}$	4·$\frac{1}{2}$	6.36	14
4·$\frac{1}{2}$	4·$\frac{3}{4}$	6.73	14
4·$\frac{3}{4}$	5	7.80	14
5	5·$\frac{1}{4}$	8.20	14
5·$\frac{3}{16}$	5·$\frac{1}{2}$	8.62	14
5·$\frac{5}{8}$	6	10.46	14
6·$\frac{1}{4}$	6·$\frac{5}{8}$	11.58	14
6·$\frac{5}{8}$	7	12.34	14
7·$\frac{1}{4}$	7·$\frac{5}{8}$	13.55	14
7·$\frac{5}{8}$	8	15.41	11·$\frac{1}{2}$
8·$\frac{1}{4}$	8·$\frac{5}{8}$	16.07	11·$\frac{1}{2}$
8·$\frac{5}{8}$	9	17.60	11·$\frac{1}{2}$
9·$\frac{5}{8}$	10	21.90	11·$\frac{1}{2}$
10·$\frac{5}{8}$	11	26.72	11·$\frac{1}{2}$
11·$\frac{5}{8}$	12	30.35	11·$\frac{1}{2}$
12·$\frac{1}{8}$	13	33.78	11·$\frac{1}{2}$
13·$\frac{1}{2}$	14	42.02	11·$\frac{1}{2}$
14·$\frac{1}{2}$	15	47.66	11·$\frac{1}{2}$
15·$\frac{1}{2}$	16	51.47	11·$\frac{1}{2}$

— List of Well Casing. —

Butt-Welded

Nominal Inside Diameter — Inches —	Actual Outside Diameter — Inches —	Thickness — Inches —	Nominal Weight per Foot — Lbs —	Number of Threads per Inch of Screw
1/8	.4	.068	.24	27
1/4	.54	.088	.42	18
3/8	.67	.091	.56	18
1/2	.84	.109	.84	14
3/4	1.05	.113	1.12	14
1	1.31	.134	1.67	11 1/2
1 1/4	1.66	.140	2.24	11 1/2

Lap-Welded

Nominal Inside Diameter — Inches —	Actual Outside Diameter — Inches —	Thickness — Inches —	Nominal Weight per Foot — Lbs —	Number of Threads per Inch of Screw
1 1/2	1.9	.145	2.68	11 1/2
2	2.37	.154	3.61	11 1/2
2 1/2	2.87	.204	5.74	8
3	3.5	.217	7.54	8
3 1/2	4	.226	9	8
4	4.5	.237	10.66	8
4 1/2	5	.247	12.34	8
5	5.56	.259	14.50	8
6	6.62	.280	18.76	8
7	7.62	.301	23.27	8
8	8.62	.322	28.18	8
9	9.68	.344	33.70	8
10	10.75	.366	40	8
11	11.75	.375	45	8
12	12.75	.375	49	8
13	14	.375	54	8
14	15	.375	58	8
15	16	.375	62	8

— List of Steam, Air and Water Pipes. —

INDEX.

	PAGE
Air Compressors—General Remarks	81
Air Engines	38
Air Engines, Exhaust Temperatures and Reheating	50
Air Receivers	182
Available Work at Complete Expansion, Curve of	30
Available Work at Full Expansion, Table of	31
Blacksmith Tools	186
Capacity of Compressors	84
Calculation of H. P. of Water, Table of	207
Circumferences and Areas of Circles	191
Column Mountings	180
Compressed Air, General Principles	3

COMPRESSORS—

Combined Duplex Steam Actuated and Shaft-driven Compressors, Class G	105
Compound Corliss Actuated Compressors, Class J	116
Direct Acting Steam Actuated Duplex Compressors, Light Duty, Class L	119
Duplex Steam Actuated Compressors, Class A	86
Duplex Shaft-driven Compressors, Class D	95
Duplex Tandem Sectional Shaft-driven Compressors, Class E	99
Light Duty Compressor or Vacuum Pump, Class K	117
Single Steam Compressors, Class B	91
Single Steam Actuated Compressors, Self-contained Type, Class C	93
Single Shaft-driven Compressors, Class F	102
Single Corliss Actuated Compressors, Class I	113
Steam Actuated Vertical Compressors, Class H	108
Steam Actuated Single Air Compressors, Class M	121

Consumption of Air	40
Consumption of Air, Single Cylinder Automatic Engines	41
Consumption of Air, Compound Automatic Engines	42
Consumption of Air, Single Cylinder Corliss Engines	43
Consumption of Air, Compound Corliss Engines	44
Consumption of Air, Corliss Compound Pneumatic Motors, Table of	51
Consumption of Air for pumps	61
Consumption of Air, Rock Drills	85
Decimals of a Foot per Each Sixty-fourth of an Inch	202
Decimals of an Inch for Each Sixty-fourth	206
Difference of Level in Use of Compressed Air	35
Difference of Level in Use of Compressed Air, Table of	37
Equation of Pipes	212
Expansion of Air	45

INDEX.

	PAGE
Fifth Roots and Fifth Powers	194
General Hints	178
Hyperbolic Logarithms	213
Indicated Horse Power to Compress Air	56
Indicated Horse Power to Compress Air, Curve for	57
Loss of Pressure in Pipes	21
Loss of Pressure through Bends	33
Loss of Pressure through Bends, Table of	34
Loss of Head in Pipes by Friction	208
Lubricators and Lubricants	187
Mean and Terminal Pressures of Compressed Air	217
Mean Absolute Pressures for Various Degrees of Isothermal Expansion, Table of	218
Mean Effective Pressures, Curve of	58
Pneumatic Governors	127
Pneumatic Hoist	52
Pneumatic Locomotive	70
Pneumatic Plant at Grass Valley	134
Pneumatic Torpedo Plant at Presidio, S. F.	140
Power Transmission by Compressed Air	71
Pressure in Vertical Pipes	59
Quantity of Air Compressed per Indicated Horse Power	53
Quantity of Air Compressed per Indicated Horse Power, Curve for	54
Reheater	48
Refrigeration by Compressed Air.	63
Rix Patent Hose Couplings	186
ROCK DRILLS—	
Rock Drills	158
Rock Drills, Rix, Table of	170
Rock Drills, Giant, Table of	171
Rock Drills, Plug and Feather	176
Sheet Iron Hydraulic Pipe, Table of	210
Squares, Cubes, and Reciprocals	195
Tripod Mountings	165
Temperatures, Centigrade, and Fahrenheit	200
Volume, Density, and Pressure of Air at Various Temperatures	215
Volumes, Mean Pressure per Stroke, Temperatures	216
Well Casing, List of	219
Wrought Iron Pipe, List of	220

www.ingramcontent.com/pod-product-compliance
Lightning Source LLC
Chambersburg PA
CBHW031811230426
43669CB00009B/1098